D1020011

Simulation
Made Easy

Simulation Made Easy

A MANAGER'S GUIDE

Charles Harrell, Ph.D.
& Kerim Tumay

Engineering & Management Press
25 Technology Park / Norcross, Georgia 30092

Printed in the United States of America

99 98 5 4 3

Library of Congress Cataloging-in-Publications Data

Harrell, Charles, 1950–
Simulation made easy : a manager's guide / by Charles Harrell and Kerim Tumay
p. cm.
Includes bibliographical references and index.
ISBN 0–89806–136–9
1. Production management—Computer simulation. 2. Stochastic processes. I. Tumay, Kerim, 1959– II. Title.
TS155.6.T84 1994
658.5'01'13—dc20 94–38721
CIP
rev.

Publisher: Ellen Snodgrass
Editor: Maura Reeves
Cover Design: Candace J. Magee
Copyedit and Composition: Jack Donner

Quantity discounts available from
Customer Service
Institute of Industrial Engineers
25 Technology Park/Atlanta
Norcross, Georgia 30092 U.S.A.
(770) 449–0460 (phone)
(770) 263–8532 (fax)

Contents

Preface ix

Foreword xi

One: **A NEW SOLUTION TO A NEW CHALLENGE** 1
- Introduction 1
- The Challenge of System Decision Making 1
- Going Upstream Without a Paddle 4
- The Simulation Solution 5
- Increasing Popularity of Simulation 6
- Uses of Simulation 7
- What Simulation Is Not 11
- When to Simulate 12
- Economic Justification of Simulation 13
- Precautions in Using Simulation 15
- Summary 15

Two: **SYSTEMS ANALYSIS AND DESIGN** 17
- Introduction 17
- What is a System? 17
- Systems Versus Processes 18
- System Elements 18
- System Performance Measures 20
- Adopting a Systems Approach 21
- System Optimization 22
- Models 24
- Types of Models 26
- Selecting the Appropriate Model 30
- Summary 31

Three: **SIMULATION BASICS** 33
Introduction 33
What is Simulation? 33
Discrete-Event Versus Continuous Simulation 34
Stochastic Versus Deterministic Simulation 35
Simulating Probabilistic Outcomes 36
History of Simulation 39
Components of Simulation Software 40
How Discrete-Event Simulation Works 44
Simulation Example 46
Summary 50

Four: **GETTING STARTED WITH SIMULATION** 51
Introduction 51
Management Commitment 51
Steps for Getting Started 52
Selecting Simulation Software 58
A Weighted Criteria Approach to Software Selection 67
Summary 72

Five: **STEPS FOR DOING SIMULATION** 73
Introduction 73
General Procedure 73
Identifying Objectives and Constraints 75
Data Collection and Analysis 77
Building An Accurate and Useful Model 85
Conducting Simulation Experiments 88
Documenting and Presenting Results 91
Pitfalls in Simulation 92
Summary 92

Six: **MODEL BUILDING** 93
Introduction 93
Modeling Paradigms 93
Abstracting System Elements 94
Modeling Entities 95
Modeling Resources 96
Modeling Movement 100
Modeling Entity Routings 103
Modeling Entity Processes 105
Modeling Entity Arrivals 108
Modeling Resource Availability Schedules 109
Modeling Resource Setups 109
Modeling Resource Downtimes and Repairs 110
Modeling Special Decision Logic 113
Summary 114

Seven: **OUTPUT ANALYSIS** 115
Introduction 115
Simulation Experiments 116
Types of Output Reports 116
Observational Versus Time-Weighted Output 117

Output Measures 119
Statistical Analysis of Simulation Output 121
Statistical Problems with Simulation Output 124
Steady-State Versus Transient Behavior 125
Terminating Versus Non-Terminating Analysis of Simulation 126
Terminating Analysis of Simulations 128
Analysis of Steady-State Behavior 132
Comparing Alternative Systems 132
Use of Random Streams 132
Summary 134

Eight: **MODELING MANUFACTURING SYSTEMS** **135**
Introduction 135
Applications of Simulation in Manufacturing 135
Decision Horizons for Manufacturing 137
Emulation 139
Manufacturing Terminology 139
Performance Measures 142
Modeling Considerations 144
Manufacturing Decision Variables 144
Types of Manufacturing Systems 153
Project Shop 154
Job Shop 157
Cellular Manufacturing 161
Flexible Manufacturing Systems 165
Batch Flow Shop 169
Line Flow Manufacturing (Production/Assembly Lines) 171
Line Flow Manufacturing (Transfer Lines) 174
Continuous Process Systems 177
Summary 178

Nine: **MODELING MATERIAL HANDLING SYSTEMS** **179**
Introduction 179
Material Handling Principles 179
Material Handling Classification 181
Conveyors 182
Industrial Vehicles 189
Automated Storage/Retrieval Systems (AS/RS) 192
Carousels 196
Automated Guided Vehicle Systems (AGVS) 197
Cranes and Hoists 202
Robots 203
Summary 205

Ten: **MODELING SERVICE SYSTEMS** **207**
Introduction 207
Applications of Simulation in Service Systems 208
Performance Measures 209
Modeling Considerations 211
General Simulation Procedures 212
Service Decision Variables 213
Types of Service Systems 215

Service Factory 216
Pure Service Shop 218
Retail Service Stores 222
Professional Services 225
Telephonic Services 227
Delivery Services 230
Transportation Services 232
Summary 235

Eleven: **MANUFACTURING AND MATERIAL HANDLING APPLICATIONS** . . 237
Introduction 237
Throughput Analysis For Aircraft Maintenance 238
Design of a Brake Valve Assembly and Test System 240
Changing from Batch Production to Continuous Flow Manufacturing 242
Multiplant Manufacturing and Distribution System 246
Cellular Design for Ophthalmic Lens Manufacturing 250
An SMT Production Line Design 252
Equipment Design and Justification 254
Optimizing Picking Methods in a Warehouse 257
Design of a Continuous Process Facility 259
Minimum Cost Test Planning 262
Business Process Re-engineering Project 264

Twelve: **SERVICE APPLICATIONS** 267
Introduction 267
Dental Clinic 267
Expansion of an Emergency Department 270
Capacity Analysis of the Lettershop Operations in a Bank 273
Client Services Help Desk 275
Design of a Fast-Food Restaurant 280

Appendix A: **Example of Input Data Collection** 283

Appendix B: **Standard Theoretical Probability Distributions** 287

Appendix C: **Confidence Interval Estimation** 294

References 297

Index 303

About the Authors 311

Preface

Whether you are a manager investigating the use of simulation as a decision making tool, an engineer or planner interested in learning how to do simulation, or a seasoned simulation practitioner, you can benefit from this book. To help you maximize the value of your reading, we would like to provide some recommendations on how to read and use the book.

The first five chapters of the book cover the fundamentals of simulation. It is important that everyone who is either directly or indirectly involved in the use of simulation understand the basics. Chapters 1 and 2 cover general topics such as the use of simulation, its benefits and drawbacks, and how it fits into the overall design of systems. Chapter 3 describes how simulation works and illustrates it with a manual simulation example. Chapter 4 contains valuable advice and practical tips on getting started with simulation and selecting simulation software. Chapter 5 explains the steps for conducting successful simulation studies and describes common pitfalls encountered in doing simulation.

If you are new to simulation, you will get a quick and practical understanding of simulation basics by reading Chapters 1 through 5. If you have an intermediate background level in simulation, you will find the first five chapters a good review. If you are an expert, you will find the material in these chapters useful for explaining to others how to get started in simulation.

Chapters 6 through 10 provide practical information for model building and output analysis that are needed to perform simulations of manufacturing and service systems. Chapter 6 discusses general system characteristics and how those characteristics are translated into a simulation model. Chapter 7 explains how to conduct simulation experiments and analyze output. Chapters 8, 9, and 10 present considerations for simulating manufacturing, material handling and service systems respectively. If you are new to simulation, Chapters 6 through 10 will provide you with a healthy dose of appreciation for the "know how" involved in performing simulation. If you are an intermediate or advanced simulation user, you will find the practical tips and techniques to be a handy reference.

Chapters 11 and 12 include numerous successful applications of simulation in manufacturing and service systems. Each application summary contains a standard format with the problem background, the objective statement, the model description, and the results. Flow diagrams and model layouts provide graphical depictions of the processes and the systems modeled in each application. Whether you are a beginner, intermediate, or advanced user, these application summaries will help you relate the modeling and analysis techniques to real world applications. They will also give you various references that you can use to educate others on the applications of simulation.

Foreword

It has been said that nothing is more powerful than an idea whose time has come. With the increasing challenge confronting businesses to develop better systems for providing goods and services, and the advent of powerful computing technology, computer simulation is emerging as a powerful decision making tool for achieving significant improvements in the design and operation of manufacturing and service systems.

Imagine being in a highly competitive industry and managing a manufacturing or service facility that is burdened by outdated technologies and inefficient work practices. To stay competitive, you know that changes must be made, but you are not sure what changes would work best, or even if certain changes will work at all. You would like to try out a few ideas, but recognize that this would be very time consuming, expensive, and disruptive to the current operation. Suppose that there was some magical way you could make a duplicate of your system and have unlimited freedom to rearrange any activity, reallocate any resource, or change any control procedure. What if you could even try out completely new technologies and methods of operation. Suppose that all of this experimentation could be done in compressed time with automatic tracking and reporting of key performance measures. Not only would you discover ways to improve your operation, but it could all be achieved risk free—without committing any capital, wasting any time, or disrupting the actual system. This is precisely the kind of capability that simulation provides. Simulation lets you experiment with a computer model of your system in compressed time giving you decision making capability that is unattainable in any other way.

Simulation is fast moving from an obscure, highly sophisticated systems analysis tool used by trained specialists to a widely used, easily understood planning tool used by engineers and managers. The most effective decision tools are those that can be used by the decision maker. Often improvements to a system suggest themselves in the very activity of building the model that an industrial engineer or operations manager would never discover if someone else is doing the model building. While an engineer or manager may not be as capable or as proficient as a "simulation expert," the dividends in increased understanding of the system operation and acquired skill in conceptualizing and analyzing system designs will pay off in the long run.

In our experience with simulation both in industry and academia over the past sixteen years, we have discovered that, despite the growth and acknowledged benefits of simulation, many engineers and operations managers in manufacturing and service industries have been reluctant to apply simulation to achieve their design and management objectives because of "lack of resources," "lack of data," or "lack of time and money." After their first simulation project, many sheepishly admitted that the real reasons for their reluctance were "lack of management awareness" and the "fear of failure" for lack of simulation "know how."

We discovered that the reason engineers and managers found simulation to be so intimidating was because they harbored misconceptions about simulation as being a difficult, theoretical tool that lacked practical application. Yes, although simulation has been taught in academic institutions, written about in engineering publications, and presented at conferences for over twenty-five years, it continues to remain somewhat of an enigma. We decided to write this book with the hope of removing the shroud of mystery surrounding simulation by providing a practical, "how to" book for managers and engineers who have been too skeptical and perhaps even timid about putting it to use.

The emphasis of the book on the practical aspects of simulation technology is in no way intended to minimize the importance of the more theoretical aspects of simulation modeling. We believe that there are many fine texts that have been written on the application of statistical methods and optimization techniques to simulation modeling and analysis. The reader is strongly encouraged to peruse one of the many excellent books available. The goal of this book is to help promote the transfer of simulation technology from the theoreticians into the hands of the practitioners.

Unlike many books on simulation which are heavily product oriented, no specific simulation language is advocated in this book. The emphasis is placed on modeling concepts and simulation procedures rather than simulation programs. Consequently, most of the models are illustrated using flow diagrams which capture the logic of the models, much the same way as logic diagrams or other systems analysis techniques do in programming. In presenting modeling methods, we have made every effort to keep the language as generic as possible. Where a decision on terminology had to be made, we used PROMODEL naming conventions since that is the product with which we are most familiar and which utilizes a natural-based language.

This book focuses on the two most common classes of systems for producing goods and services; namely, manufacturing and service systems. Nearly 15 percent of the U.S. workforce is employed in manufacturing. In 1955, about one-half of the U.S. workforce worked in the service sector. Today, nearly 80 percent of the U.S. workforce can be found in service-related occupations. Manufacturing and service systems share much in common. They both consist of activities, resources, and controls for processing incoming entities. The performance objectives in both instances relate to quality, efficiency, cost reduction, process time reduction, and customer satisfaction. In addition to having common elements and objectives, they are also often interrelated. Manufacturing systems are supported by service activities such as product design, order management, or maintenance. Service systems receive support from production activities such as food production, check processing, or printing. Regardless of the industry in which one is involved, an understanding of the modeling issues underlying both systems will be helpful.

The book will be of greatest value to decision makers such as engineering managers, operations managers, directors, vice presidents, general managers, presidents, and CEOs. These individuals have neither the time nor the interest in doing simulation themselves. Yet, they are the drivers of change in their business. They will find the book an excellent source of information for "championing" simulation and seeing that it is used as a business tool to increase profitability, reduce costs, and maintain competitive advantage.

The book will be also be of value to simulation practitioners and consultants in manufacturing and service organizations. Industrial engineers, management engineers, project engineers, manufacturing engineers, process engineers, equipment engineers, sales engineers, and quality engineers will find the book a convenient reference for modeling and analysis purposes.

Although this book is not written as a textbook for academic use in a "traditional" simulation course, those educators who are interested in teaching the practical applications of simulation in the design and management of manufacturing and service systems may find the book appropriate for their students.

Acknowledgments

We would like to recognize and acknowledge Maura Reeves, senior editor at Industrial Engineering and Management Press, for the guidance and support she gave us throughout the entire project. This made it possible for us to complete the book in a timely fashion.

The last two chapters of the book were made possible by contributions from many individuals. We would specifically like to thank Kal Mabrouk, William Reedy, Ken Davis, Ali Kiran, Roy Henderson, Steve Strudler, Roland Schelasin, John Mauer, Rob Singer, Craig Lewis, Jerry Hein, Tom Jones, Mike Kraitsik, Suzanne Whitworth, Craig Marmon, Dan Roberts, Carmen Guerrera, Ed Nixon, Anthony Gasparatos, and Pete Nedza for providing us with their original papers, and necessary information and graphics for this book.

Special thanks go to Charles White, Kal Mabrouk, Ali Kiran, Steve Courtney, and Kunter Akbay for reviewing the book and providing us with valuable feedback for improvements.

This book was written with the support of PROMODEL Corporation and Brigham Young University. We are indebted particularly to Scott Baird, president and CEO of PROMODEL, for encouragement and support during the writing of this book. Many other members of PROMODEL provided valuable input and suggestions.

Most importantly, we express our thanks to our devoted wives, Yvonne Harrell and Susan Tumay, for their enduring patience and support throughout the development of this book.

Chapter 1
Simulation: A New Solution to a New Challenge

"It must be remembered that there is nothing more difficult to plan, more doubtful of success, nor more dangerous to manage, than the creation of a new system."

Niccolo Machiavelli

INTRODUCTION

Machiavelli's statement five hundred years ago regarding the challenge of planning and managing political systems is equally applicable to the design and operation of modern day manufacturing and service systems. These systems, which consist of resources, activities, and controls for providing goods and services, have shorter life cycles, greater complexity, and higher performance requirements than ever before. To help make the kinds of system design and operation decisions that will enable a business to remain competitive, computer simulation is being successfully used in a growing number of applications.

What are the issues associated with designing and managing manufacturing and service systems? How have businesses traditionally dealt with these issues? What is simulation and how does it compare with traditional approaches to making system design and operational decisions? How can the cost of simulation be justified? Are there drawbacks to using simulation? This chapter addresses these questions and provides a look into this powerful new way of making the challenging decisions required of today's manufacturing and service systems.

THE CHALLENGE OF SYSTEM DECISION MAKING

Fierce global competition, rising consumer demands, and advancing technologies have all combined to force companies to rethink the way they are doing business. Today's customers demand high quality, customized goods and services at low prices — and they don't want to wait. As customer expectations continue to increase with respect to quality, customization, price and delivery, businesses are paying greater attention to the *way* goods and services are being provided, and not just *what* goods and services are provided. The push to more effectively integrate and streamline manufacturing and service operations has intensified, as characterized by social philosopher Alvin Toffler (1980):

> Vast changes in the technosphere and the infosphere have converged to change how we make goods. We are moving beyond traditional mass production to a sophisticated mix of mass and demassified products. The ultimate goal of this effort is now apparent: complete customized goods made with holistic, continuous flow processes.

At a time when the focus on improving production and service systems is greater than ever before, there has never been a time when the design and management of these systems has been more challenging. This challenge has been brought on by the following factors:

- Systems have *shorter life cycles* due to constantly changing requirements and the proliferation of new technologies.
- Systems are becoming increasingly *more complex* as a result of growing technology sophistication and greater process integration.
- Systems have *higher performance requirements* due to increased competition and rising customer expectations.

These trends are shown in Figure 1.1.

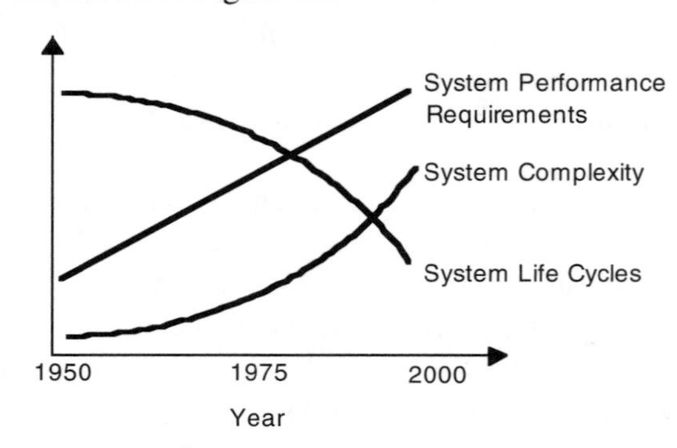

Figure 1.1 Increasing System Performance Requirements, Growing Complexity, and Shorter Life Cycles.

Shorter System Life Cycles

Living in an era where the only thing that is constant is change, it is only natural to expect to see manufacturing and service systems under constant revision. Some of the reasons why system changes become necessary include the following:

- Over time, equipment begins to show signs of wear and failure and requires replacement.
- New or modified products and services require new or modified systems.
- An increase or decrease in production volume or service loads requires system adjustments.
- Increased competition or declining profit makes it necessary to find ways to operate more efficiently.
- New technology makes old ways of doing things obsolete.

With the growing increase in the frequency of system changes, system life cycles are becoming shorter and shorter. Where production of the Model T Ford continued for an entire generation, a production setup today rarely extends beyond a couple of years. By the time a new system has been put in place, processing requirements have already changed and newer, improved technologies have made existing ones obsolete. Conway and Maxwell (1986) describe this situation as it relates to manufacturing: "We no longer have the luxury of time to tune and debug new manufacturing systems on the floor, since the expected economic life of a new system, before major revision will be required, has become frighteningly short." Businesses that are succeeding are those that are able to keep up with this rapid pace of change.

Increasing System Complexity

As a result of advancing technologies and the growing need to modernize and streamline manufacturing and service systems, systems planning has become increasingly more complex. Not only are the issues involving the selection, configuration, and integration of these technologies becoming more complicated and confusing, but it is also difficult to predict exactly how much improvement, if any, will be gained by incorporating a particular technology.

In addition to becoming more complex, systems are expanding to encompass more of the activities related to the operation. Companies are now integrating forward to include customers, and backward to interact more closely with suppliers. This trend toward expanding business operations makes the design and management of systems much more challenging.

Increasing Performance Requirements

On top of the challenge of dealing with increasing system complexity and decreasing system life cycles, businesses are now under greater pressure than ever before to maximize the operational efficiency of systems. Inefficient operating practices are all around us. It is not difficult to walk through a factory or visit a hospital and recognize one or more activities that could be easily improved through minor changes. Excessive waiting times, unnecessary redundancies, and other non-value-added activities plague nearly every business. Value-added time for manufacturing industries, for example, often comprises only three to five percent of the total throughput time. IBM Credit Corporation discovered that value-added time for the financing approval process amounted to only ninety minutes out of a three- to four-day cycle time. Several insurance companies have been observed to spend only two or three hours in an overall twenty-day cycle actually processing and underwriting the insurance policy (Davenport 1993).

Inefficiencies and other forms of waste are frequently subtle and difficult to detect. One company, for example, thought that running large lot sizes was a good solution to dealing with long setup times. It had not occurred to them that it might be possible to reduce the setup time until an outside consultant pointed out the solution. As a result of implementing a simple swivel table from which tooling could immediately be changed, they were able to cut setup time from two hours down to ten minutes. Over a period of years, the erosion of profits caused by such wasteful practices can be devastating. "The trick is to find waste, or *muda*" advises Shingo (1992). "After all, the most damaging kind of waste is the waste we don't recognize."

GOING UPSTREAM WITHOUT A PADDLE

With the challenge facing companies to make quicker, more difficult decisions that have a greater impact on the design and operation of manufacturing and service systems, engineers and managers are desperately seeking better tools to aid in systems design and operational planning. Traditional methods, such as work analysis, flow charting, process mapping, linear programming, etc. are incapable of solving the complex integration problems of today. These tools have only limited application and are unable to provide a reliable measure of expected system performance.

In the absence of useful planning and evaluation tools, most companies resort to trial-and-error methods which end up being costly, time consuming and too disruptive to provide any useful benefit. Solberg (1988) notes:

> The ability to apply trial-and-error learning to tune the performance of manufacturing systems becomes almost useless in an environment in which changes occur faster than the lessons can be learned. There is now a greater need for formal predictive methodology based on understanding of cause and effect.

Trial-and-error methods are based only on a hunch both about what solution will work best and how much improvement a given solution can actually provide. This trial-and-error method of system improvement is much like playing Russian roulette — no one is ever certain what the outcome is going to be and, if a wrong decision is made, the consequences can be fatal to a business.

THE SIMULATION SOLUTION

One tool that is rapidly gaining popularity in systems design and analysis is computer simulation. Simulation is a powerful analysis tool that helps engineers and planners make intelligent and timely decisions in the design and operation of a system. Simulation itself does not solve problems, but it does clearly identify problems and quantitatively evaluate alternative solutions. As a tool for doing "what if" analysis, simulation can provide quantitative measures on any number of proposed solutions to help quickly narrow in on the best alternative solution. By using a computer to model a system before it is built or to test operating policies before they are actually implemented, many of the pitfalls that are often encountered in the startup of a new system can be avoided. Improvements that previously took months and even years of fine tuning to achieve can now be attained in a matter of days or only hours through computer simulation.

The ability of simulation to consider a large number and wide variety of variables in a single model makes it an indispensable tool in designing today's complex business systems. Kochan (1986) notes that in manufacturing systems:

> The possible permutations and combinations of workpieces, tools, pallets, transport vehicles, transport routes, operations, etc., and their resulting performance, are almost endless. Computer simulation has become an absolute necessity in the design of practical systems, and the trend toward broadening its capabilities is continuing as systems move to encompass more and more of the factory.

For service systems, the complexities of customer scheduling, staffing, resource management, customer flow and information processing have made simulation equally valuable.

The benefits of simulation are much like those realized in a flight simulator in which a pilot gains experience without the risk, time, and cost associated with training on actual equipment. In flight simulation, a pilot's skills are sharpened and his performance improved through interacting with the simulator. The pilot is allowed to make mistakes in a simulated environment so that errors are reduced to a minimum in actual flight. Like flight simulation, simulation of manufacturing and service systems is performed to improve one's understanding of how the system operates so that intelligent and skilled decisions can be made, to reduce the time and cost associated with experimenting on the real system, and to minimize the risk of making mistakes on the actual system.

INCREASING POPULARITY OF SIMULATION

Simulation was first used in the 1950s, primarily in military strategic planning. It has only been in the last decade, however, that it has gained popularity in manufacturing and service industries. For many companies, simulation has become a standard practice when a new facility is being planned or a process change is being evaluated. Surveys indicate that simulation ranks first among leading management science and operations research techniques in terms of popularity and usefulness (Shannon 1980).

Several factors have contributed to the increased use of simulation including the following:

- Increased awareness and understanding of the technology.
- Increased availability, capability and ease-of-use of simulation software.
- Increased computer memory and processing speeds (especially at the PC level).
- Declining computer hardware costs.
- Widespread adoption of microcomputers in businesses.

Simulation is no longer considered to be a method of "last resort." The availability of easy-to-use simulation software and powerful desktop computers have not only made simulation more acceptable, but have made simulation more accessible to the designer or manager who has neither the time nor the interest to learn difficult and complicated analysis techniques. Simulation is becoming to systems planners what spreadsheets have become to accountants.

Simulation is being used in a variety of manufacturing and service industries. A list of some of the industries that are currently benefiting from the use of simulation is given in Table 1.1. Selected examples of actual simulation studies performed in several of these industries are provided in Chapters 11 and 12 of this book.

It is of interest to note that 59 percent of all applications are in manufacturing industries (Christy and Watson 1983). Manufacturing systems tend to have clearly defined relationships and formalized procedures that lend themselves to modeling analysis. Recent trends to formalize and systematize all business processes (order processing, invoicing, customer support, etc.) through standardizing techniques, such as those specified in ISO 9000, have made simulation a well suited tool for other business processes as well. It has been observed that 80 percent of all business processes are repetitive processes that can benefit from the same analysis techniques used to improve manufacturing systems (Harrington 1991). With this in mind, the use of simulation in designing and improving business processes of every kind is likely to continue to grow.

Table 1.1 Sample Manufacturing and Service Industries Where Simulation is Used.

Manufacturing Industries	Service Industries
Appliance	Public services
Automotive	Learning institutions
Aerospace	Restaurants and fast food
Electronics	Banking
Heavy equipment	Health care
Glass and ceramics	Government
Clothing apparel	Disaster planning
Rubber and plastics	Waste management
Food and beverage	Transportation
Mills (steel, paper, etc.)	Distribution
Foundries	Aerospace-military
Petrochemical	Hotel management
Furniture	Amusement parks

USES OF SIMULATION

Simulation is a versatile tool that has been used in a number of different ways including the following:

- Systems design
- Systems management
- Training and education
- Communication and sales
- Public relations

Below is a description of how simulation has been found to be useful in each of these areas.

Systems Design

In designing a *new* system, experiments can be performed on a model that would otherwise be impossible to carry out on the actual system since it is not yet implemented. Simulation can significantly reduce the time required to debug and fine-tune the system once it is installed.

In making improvements to an *existing* system, simulation allows experimentation to be performed on a model so that the actual system need not be disturbed. Working with imaginary objects and resources provides much more flexibility in making changes and is much less costly than experimenting with the actual system.

When used primarily as a design tool, whether in designing a new system or modifying an existing system, simulation addresses such areas as:

- *Methods selection.* Should several activities all be performed at a single station or broken up into several operations?
- *Technology selection.* What is the effect of using automation instead of manual processing?
- *Optimization.* What is the optimum number of resources that best achieves performance objectives?
- *Capacity analysis.* What is the throughput capacity of the system?
- *Control system decisions.* Which tasks should be assigned to which resources?

As a systems analysis tool, simulation is to systems design what engineering analysis is to product design, or what inspection is to a manufacturing operation. That is, it is a testing tool to be applied as early as possible in order to catch mistakes before they become costly. An important rule of production, called the *rule of tens* and credited to Dr. Ohno of Toyota, states that "an error that slips through increases in cost to remedy by a factor of ten with each operation until it is detected." This rule can be modified to apply to systems planning as follows: "The cost to make improvements to a system increases tenfold with each additional stage of progress" (see Figure 1.2). The idea behind simulation is to make as many of the changes as possible during the early concept and design stages of a system where the cost of making changes is kept to a minimum.

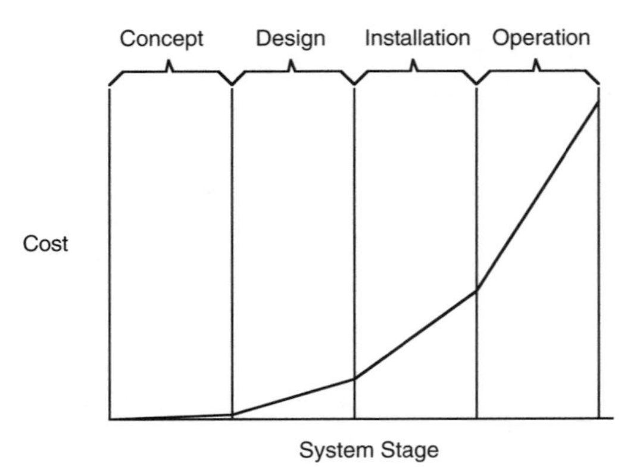

Figure 1.2 Cost of Making Changes at Each Stage of System Development.

Even if no problems are uncovered through simulation during the design phase, the exercise of developing a simulation model of the system is beneficial from the standpoint of providing a systematic, structured approach to ensure that all of the operational issues of the system have been properly addressed. It also enables one to gain an overall understanding of the system dynamics that would otherwise be difficult to obtain.

To develop a system design that will work, it is not enough to just define *what* tasks will be performed and *what* resources will be needed. The mechanics must also be worked out specifying *how* the work will be done. For lack of a way to focus on operational issues during the design phase, these details are often overlooked. The philosopher Alfred North Whitehead pointed out this tendency of the human mind to gloss over details when he stated, "We think in generalities, we live in detail." The world is full of generalists who are quick to give broad brush solutions but slow to produce specific, workable implementations. Simulation makes certain that the operational details of a system are addressed in the early design stages so that questions that arise later have already been resolved.

The increased discipline that simulation brings to the design process together with the capability of taking a total systems approach to design are sometimes sufficient reasons in themselves for employing simulation. Often improvements are discovered solely as a result of going through the model building process — before any simulation run is made.

Systems Management

In managing the operation of a system, simulation helps determine the best way of controlling the flow of customers and materials. It also helps in finding the most effective way to schedule and deploy resources. Simulation replaces the wasteful and often unreliable practice of setting management policies based on trial-and-error methods. Deming (1989) states: "Management of a system is action based on prediction. Rational prediction requires systematic learning and comparisons of predictions of short-term and long-term results from possible alternative courses of action." The key to sound management decisions lies in the ability to accurately predict the outcome of alternative courses of action. Simulation provides precisely that clarity of foresight. By simulating alternative production schedules, operating policies, staffing levels, job priorities, decision rules, etc., a manager can more accurately predict outcomes and therefore make more intelligent and informed management decisions.

In systems management, simulation assists in making the following decisions:

- *Production/Customer Scheduling.* What is the best sequence and timing for introducing products or admitting customers into the system?
- *Resource Scheduling.* What personnel and equipment are needed on which shifts?
- *Maintenance Scheduling.* What preventive maintenance schedule is the least disruptive to the system operation?

- *Work Prioritizing.* What is the best way of prioritizing tasks to maximize efforts?
- *Flow Management.* What is the best way to keep the flow of materials/customers spread evenly in the system?
- *Delay/Inventory Management.* What is the most effective way to keep customer waiting or inventory levels to a minimum?
- *Quality Management.* How will operations be impacted if inspection points are eliminated and personnel assume full responsibility for the quality of their work?

Training and Education

For training and education, since simulation contains only the driving elements of a system, it provides a clear insight into the system dynamics. Simulation can help operators, service representatives, or supervisors learn how a system operates, and what happens when alternative management decisions or operating policies are implemented. This prepares operating personnel to be more effective when working with the actual system.

Simulation is now widely used in educational settings to help students understand the complex interactions that exist in different service and manufacturing systems. It allows them to try out problem solving techniques learned in the classroom to see which methods work best under different situations. Through several hours of interacting with a variety of simulation models, a student can gain an understanding of the issues and driving forces behind a number of different types of systems that would otherwise take years of experience to obtain.

A relatively new and promising simulation method used for education and training is *interactive modeling* in which students interact with a simulation while it is running. This allows students or management trainees to acquire skills and confidence in making real-time management decisions that will be of valuable benefit later when working with an actual system. Equipment failures, quality problems, employee absenteeism, etc. can all be incorporated to make a "true to life" system model. Participants can work individually or in teams to try their skills at dealing with the operational issues associated with different business systems.

Communication and Sales

In communicating and selling system innovations, the animation displayed during simulation provides an excellent visual aid for demonstrating how the system will perform. It is surprising how effective an animated representation of a system is for getting management's attention and influencing their thinking. Individuals who previously dozed through technical presentations suddenly perk up and take notice.

An animated simulation is especially convincing in demonstrating the effectiveness of a concept when attempting to sell a particular solution to a customer or client. Even selling ideas internally can be achieved more effectively through the use of simulation. Managers are often reluctant to entertain new technologies either because of a lack of understanding or because of a lack of confidence. Simulation can help alleviate management concerns by giving them visual and quantitative proof that the system will work. Just as a single picture is worth a thousand words, a complete animation is worth a thousand pictures.

Public Relations

Businesses that adopt leading edge technologies tend to be viewed as progressive, forward thinking companies. Simulation is a highly visible technology that elevates a company's image both in the eyes of the public as well as in the eyes of its customers. It conveys a message to the outside world that the company takes pride in how it operates.

Companies in the past have displayed scale models or pictures of their facilities strategically situated in a front lobby for visitors to see. In the future, these scale models may very well be replaced by three-dimensional animations with accompanying sound that takes visitor's through a guided tour of the facility.

WHAT SIMULATION IS NOT

While simulation can be a powerful and useful tool when correctly applied, it is not a panacea for all system related problems. Simulation is primarily intended to address the operational aspects of a system—*what, when, where* and *how* tasks are performed. It goes without saying that not all of the issues affecting system performance are operational issues. There are also human and technological issues to be addressed. In fact, manufacturing and service systems are often referred to as sociotechnical systems to emphasize the important role that both human and technological elements play in the operation of the system. It is the combination of these three aspects of a system (human, technological and operational) that ultimately determines system performance (see Figure 1.3).

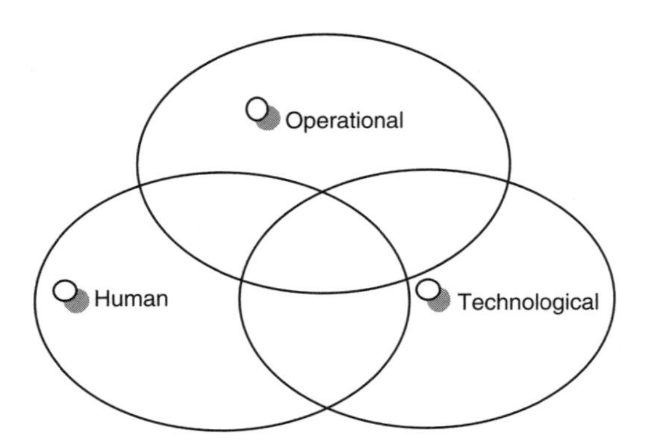

Figure 1.3 Aspects of a System that Determine Performance.

Human aspects of a system include skill levels, interests, and management/employee relations. Technological issues include equipment reliability, power requirements, and process capability. Simulation is unable to address these human and technological issues *per se*. It does, however, show how a system performs when given certain human and technological characteristics.

In addition to being confined to operational related issues, simulation is also limited to providing evaluations, not solutions. It is the modeler that must be able to generate the solutions to be evaluated. Simulation is no substitute for careful thinking and analysis. It should be viewed as an extension to the human mind that allows complex relationships to be conceptualized, and multiple factors to be simultaneously examined — something which the human mind alone is incapable of doing.

WHEN TO SIMULATE

Not all problems that *can* be solved with the aid of simulation *should* be solved using simulation. It is important that the tool fits the task. If the objective is to understand work flow sequence, then a simple flow diagram will suffice. For simple problems like finding the effective capacity of an operation, or the cumulative scrap rate of a system, simple mathematical calculations are better suited. You don't want to use a shotgun to kill a fly.

Often the complexity of the system is the main determinant for choosing to use simulation. For example, the effect of adding a second machine to an operation is easy to determine if one is only dealing with a single operation. If, however, the operation is one of several in a series of operations, the effect this change has on the overall throughput of the system is much more difficult to predict.

As a general guideline, simulation is an appropriate tool if:

- Developing a mathematical model is too difficult or perhaps even impossible.
- The system has one or more interdependent random variables.
- The system dynamics are extremely complex.
- The objective is to observe system behavior over a period of time.
- The ability to show the animation is important.

ECONOMIC JUSTIFICATION OF SIMULATION

Much of the reluctance to using simulation comes from the mistaken notion that simulation is costly and very time consuming. The cost of simulation consists of the initial investment and startup costs and the individual project modeling costs. The initial investment, including training, may be between $10,000 and $30,000. This cost is often recovered within the first project or two. The ongoing expense of using simulation is estimated to be between one and three percent of the total project cost (Glenney and Mackulak 1985). In many applications, the savings from the project far exceed the cost of the simulation. Much of the effort that goes into building the model is in arriving at a clear definition of how the system operates, which needs to be done anyway. With the advanced modeling tools that are now available, the model development and experimentation phase may only take a few days or weeks which is usually only a small fraction of the overall project development time.

Simulation helps achieve the greatest improvements in manufacturing and service system performance in the shortest amount of time. The cost and time to simulate a system becomes minuscule compared with the long-term savings from having efficiently operating systems. Figure 1.4 illustrates how the cumulative cost associated with systems designed with simulation can compare with those designed without the use of simulation. Note that while the short-term cost may be higher due to the extra time needed to optimize the system design, the long-term operating costs are much lower. Refusing to use simulation on the basis of cost is a failure to recognize the long-term savings that come from having an effectively designed and efficiently operating system.

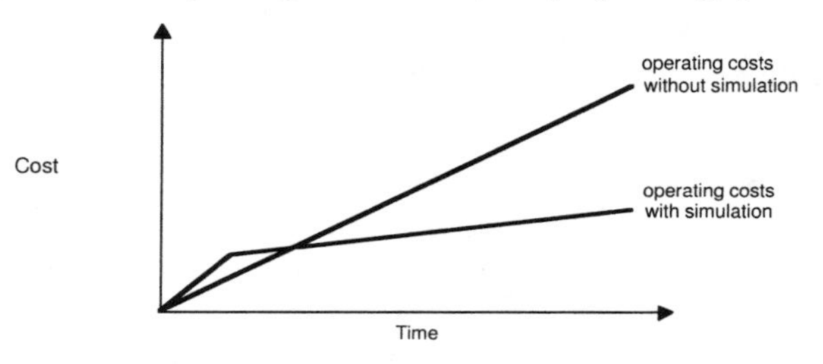

Figure 1.4 Comparison of Operating Costs With and Without Simulation.

Simulation during the design phase of a system results in cost savings by identifying and eliminating unforeseen problems and inefficiencies. Cost is also reduced by eliminating overdesign and removing excessive safety factors that are added when performance projections are uncertain. It is not uncommon for companies to report hundreds of thousands of dollars in savings through the use of simulation. One Fortune 500 company was designing a facility for producing and storing subassemblies in preparation for the final assembly of a number of different metal products. One of the decisions involved determining the number of containers required for holding the subassemblies. It was initially felt that 3,000 containers were needed to handle the activity. However, after a simulation study it was found that throughput did not significantly change when the number of containers varied between 2,250 and 3,000. By purchasing 2,250 containers instead of 3,000, a savings of $528,375 was expected in the first year with annual savings thereafter of over $200,000 due to the savings of floor space resulting from having 750 fewer containers (Law and McComas 1988).

Even if dramatic savings or improvements are not realized each time a model is built, at least it inspires confidence that a particular system design is capable of meeting required performance objectives and thus minimizes the risk often associated with new startups. The economic benefits derived from instilling confidence in a proposed system were evidenced in a situation in which a simulation consultant working with an entrepreneur developed a model of a blanket factory for which the latter was attempting to secure bank financing and investor interest. Based on the processing times and equipment lists supplied by industry experts, the model showed that the output projections in the business plan were well within the capability of the proposed facility. Although unfamiliar with the blanket business, bank officials felt more secure in agreeing to support the venture (Harrell, Bateman, Gogg, and Mott 1992).

Often, simulation can help achieve productivity improvements without the need to invest heavily in new technologies or facility expansions. By looking at the overall operation of the system in compressed time, long standing problems such as bottlenecks, redundancies and inefficiencies that previously went unnoticed start to become more apparent. Consider the following examples:

- GE Nuclear Energy was able to increase its output of highly specialized reactor parts by 80 percent. The cycle time required for production of each part was reduced by an average of 50 percent. These results were obtained by running a series of models, each one solving production problems highlighted by the previous model (Harrell, Bateman, Gogg and Mott 1992).

- A large manufacturing company with stamping plants located throughout the world produced stamped aluminum and brass parts on order according to customer specifications. Each plant had from 20 to 50 stamping presses which were utilized anywhere from 20 to 85 percent. A simulation study was conducted to experiment with possible ways of increasing capacity utilization. As a result of the study, machine utilization improved from an average of 37 to 60 percent (Hancock, Dissen, and Merten 1977).
- A diagnostic radiology department in a community hospital was modeled to evaluate patient and staff scheduling, and to assist in expansion planning over the next five years. Analysis using the simulation model enabled improvements to be discovered in operating procedures that precluded the necessity for any major expansions in department size (Perry and Baum 1976).

PRECAUTIONS IN USING SIMULATION

While simulation has been proven to be beneficial, it is not without potential dangers. Some precautions to be taken in using simulation are as follows:

- Can be expensive and time consuming to initially get started.
- Sometimes easier and better solutions get overlooked.
- Results can be misinterpreted.
- Human and technological factors may get ignored.
- May place too much confidence in simulation results.
- It is difficult to verify whether the results are valid.

While simulation technology is growing and becoming more powerful in relieving the modeler from tedious details, there still remains a great deal of effort on the part of the modeler to use the tool intelligently.

It is the goal of this book to show when and where the use of simulation is appropriate, and to explain how to use simulation in making design and management decisions.

SUMMARY

Businesses today face, more than ever before, the challenge of quickly implementing complex production and service systems that are capable of performing at maximum efficiency. With recent advances in computing and software technology, simulation tools are now available which help overcome the challenges in planning and managing manufacturing and service systems.

Simulation offers many benefits to systems managers and engineers. These benefits can be summarized as follows:

- Accounts for complex interdependencies and variability.
- Versatile enough to model almost any type of system.
- Shows performance changes over time.
- Permits controlled experimentation.
- Non-disruptive of the actual system.
- Easy to use and understand.
- Stimulates interest and team participation.
- Visually realistic and convincing.
- Forces attention to detail in a design.

Any company that is designing a new system, or seeking to make improvements to its existing system can benefit from the use of simulation. In the remainder of this book, we look at how simulation works and how to use simulation in making system design and management decisions.

Chapter 2
Systems Analysis and Design

"A fool with a tool is still a fool."

Unknown

INTRODUCTION

While this book is ostensibly about simulation, it is really a book about systems analysis and design. Simulation is, after all, merely a tool that is best understood within the context of the problems for which it is intended to help solve; namely, system operational problems. Simulation is of little benefit if the system being modeled is not understood and sound system design and improvement principles are not practiced. This chapter looks at the components that make up a system and the factors that impact system performance. The role of modeling in systems analysis and design is described with an explanation of the general concepts behind modeling. A comparison is made between symbolic modeling, analytic modeling, and simulation modeling.

WHAT IS A SYSTEM?

Broadly defined, a system is a collection of elements that function together to achieve some objective (Blanchard 1991). Examples of systems are stellar systems, ecosystems, traffic systems, political systems, economic systems, manufacturing systems and service systems. Systems may be classified in many different ways. For example, we may classify them according to the nature of their behavior (static vs. dynamic, deterministic vs. stochastic, time varying vs. time invariant) or according to the function they perform (circulatory, structural, transformational) For purposes of this book, the term *system* will be used throughout to refer to manufacturing and service systems that process discrete entities such as customers and products. An example of a manufacturing system is a production line where machines and workers convert raw materials into finished products. A service system might be an emergency room of a hospital in which nurses, doctors, and staff are used to treat incoming patients.

SYSTEMS VERSUS PROCESSES

With the recent attention being given to process reengineering and process improvement, a clarification might be helpful to understand the relationship between systems and processes. In the seminal book entitled *Reengineering the Corporation*, Hammer and Champy (1993) define a process as the collection of *activities* that create an output based on one or more inputs. A system, on the other hand, is a collection of *elements* used to perform a process. Thus, systems encompass processes but also include the resources and controls for carrying out processes.

In process design, the focus is on *what* is being performed in the system. In systems design, the emphasis is on the details of *how, where,* and *when* the process is performed. Defining the process generally comes first, followed by the design of the system. Each one, however, can have an influence on the other.

SYSTEM ELEMENTS

At the most basic level, a system is made up of *entities, activities, resources,* and *controls* (see Figure 2.1). These elements define the who, what, where, when, and how of system processing.

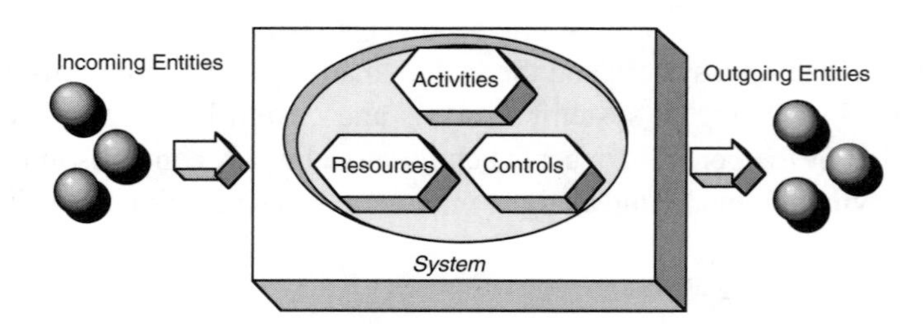

Figure 2.1 Elements of a System.

Following is a brief description of these system elements with examples of each.

Entities

Entities are the items processed through the system such as products, customers, and documents. Different entities may have different characteristics such as cost, shape, priority, status, or condition. Entities may be classified into three types:

- Human or animate (customers, patients, etc.)
- Inanimate (parts, paperwork, bins, etc.)
- Intangible (calls, electronic mail, projects, etc.)

For some systems (called continuous systems), there exists no discrete entity, but rather the input and output are a continuous flow of material, such as in an oil refinery.

Activities

Activities are the tasks or actions that take place in the system, such as filling a customer order, cutting a part, or repairing a machine. Activities have duration and usually involve the use of resources. Typical kinds of activities include:

- Entity processes (check-in, treatment, inspection, fabrication, etc.)
- Entity moves
- Resource moves
- Resource setups
- Resource maintenance and repairs

Resources

Resources are the means by which activities are performed. Resources define who or what performs the activity, and where the activity is performed. Resources may have characteristics such as capacity, speed, cycle time, and reliability. Typical resources used in a system might include the following:

- Personnel
- Equipment
- Space
- Tooling
- Energy
- Time
- Money

Controls

Controls govern how, when, and where activities are performed. They also determine what action is taken when certain events or conditions occur. At the highest level, controls are in the form of schedules, plans, and policies. At the lowest level, controls are in the form of written procedures and computer logic. Examples of controls include:

- Process plans
- Production plans
- Work schedules

- Maintenance policies
- Programmable logic controller (PLC) programs
- Limit switches
- Instruction sheets

SYSTEM PERFORMANCE MEASURES

The goal of system design and improvement efforts is to transform inputs into the desired outputs in the most timely, efficient, and cost effective manner. To measure the effectiveness of a system and to evaluate when improvements are made, it is necessary to have performance measures. Following is a description of common performance measures used to gauge the effectiveness of a manufacturing or service system:

- *Cycle time.* The throughput or service time for processing material or customers.
- *Resource utilization.* The percentage of time that equipment and personnel are in productive use.
- *Value-added time.* The amount of time that material and customers actually spend in operation or receiving service.
- *Waiting time.* The amount of time that material and customers spend waiting for operations and services.
- *Processing rate.* The throughput or service rate of material, customers, etc.
- *Quality.* The percentage of parts produced or customers served that meet a defined set of standards.
- *Cost.* The operating costs of the system.
- *Flexibility.* The ability of the system to adapt to fluctuations in volume and variety.

In establishing performance goals for a system, nebulous goals or fad phrases such as becoming a world class manufacturer or a total quality service organization should be avoided. Goals are more effective if they are specific and include a measurable objective. System performance goals should be consistent with overall company goals and should be based on internal and external benchmarking.

Actual performance goals will depend on the specific industry and nature of the operation. Examples of specific, quantitative goals for a system might be:

- Maintain resource utilizations of at least 80 percent.
- Keep in-process inventory levels to under 50 units.
- Keep annual operating cost increases under 10 percent.
- Keep waiting times under 15 minutes.

ADOPTING A SYSTEMS APPROACH

Designing a new system or making improvements to an existing system requires more than simply identifying the elements and performance goals of the system. It requires an understanding of how system elements relate to each other and to overall performance goals. We call this approach to systems design, which takes into account the interaction and interrelationships of all elements in the system, a *systems approach*. Because systems are composed of interdependent elements, it is not possible to predict how a system will perform simply by examining each of the system elements in isolation of the whole. While structurally a system may be divisible, functionally it is indivisible and therefore requires a systems approach to understand its operation.

Due to compartmentalization and specialization, system change decisions are often made without having a complete knowledge of all of the cause-and-effect relationships in the system. With everyone busy taking care of individual pieces of the system, few, if any, individuals in an organization are watching out for overall system behavior. One manager in a large manufacturing corporation we worked with noted that as high as 99 percent of the system improvement recommendations that are made in his company fail to look at the whole picture. He further estimated that nearly 80 percent of the suggested changes resulted in no improvement at all, and many of the suggestions actually hurt performance. When attempting to make system improvements, it is often discovered that localized changes fail to produce the overall improvement that is desired. The elimination of one problem area only serves to uncover, and sometimes even exacerbate, other problem areas.

Businesses that persist in viewing problems myopically will continue to make decisions that are not always in the best interest of the overall system performance. Improving system performance requires a broadening of perspective to look at the complete picture. Specifically, it requires an understanding of all relevant cause-and-effect relationships in the system as well as *key* decision-response relationships.

Cause-and-Effect Relationships

The cause-and-effect relationships in a system define the behavior or dynamics of the system and, therefore, determine how the system will perform. Cause-and-effect relationships are defined by identifying all of the actions that can take place in the system, and then determining the events, conditions, or other actions that trigger each of these actions. Some cause-and-effect relationships are inherent in the system in that the effect is a natural consequence of some action. For example, after 50 to 60 cycles on a particular machine, tool wear reaches a point where it requires replacement. Other cause-and-effect relationships are human-imposed and are specified as a planned reaction to occur as the result of some action. For example, the arrival of a customer to a service counter should cause a service representative to wait on the customer. Many cause-and-effect relationships are actually part of chain reactions in which a series of resultant actions are triggered by an initiating action.

Decision-Response Relationships

While cause-and-effect relationships deal with direct and usually immediate reactions to triggering causes, the overall performance or response of the system is the result of all of the combined effects that take place over a period of time. Overall system response is not a linkage that can be prescribed or dictated. The response of a system to given decision variable settings can only be estimated analytically or determined empirically (through experimentation).

A decision variable is the specification for a particular system element (resource quantity, activity time, control logic, etc.). A system operates as defined by decision variables which cause the system to respond a certain way. Response variables (sometimes called performance variables) are variables that measure the performance of the system in response to a particular set of decision variables.

In determining decision-response relationships, it is not necessary to identify how system performance responds to every possible decision variable. What is essential to know is how performance is impacted by key decision variables (variables that are the subject of investigation and are practical to change).

SYSTEM OPTIMIZATION

Whether designing a new system, or improving an existing system, it is important to follow sound design principles. The science of systems design, also called systems engineering, has been defined (Blanchard 1991) as:

> The effective application of scientific and engineering efforts to transform an operational need into a defined system configuration through the top-down iterative process of requirements definition, functional analysis, synthesis, optimization, design, test and evaluation.

Systems design essentially involves defining objectives, identifying requirements, specifying a solution, and then evaluating the effectiveness of the solution in meeting overall objectives (see Figure 2.2).

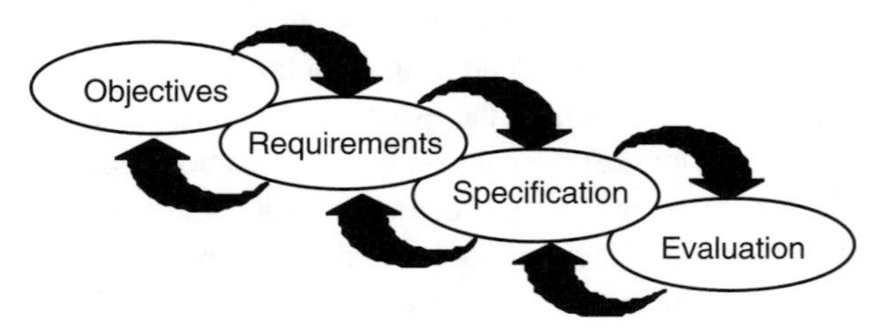

Figure 2.2 Four Phase Iterative Approach to Systems Design.

An important and common objective of many system development efforts is to achieve an optimum system configuration. Optimization efforts seek to find the best combination of system design and operating parameters (decision variable values) that either minimizes or maximizes some performance measure (the response variable value). Examples of optimization objectives might be to minimize costs or maximize productivity. The objective of the optimization effort is referred to as the *objective function.*

In some instances, we may find ourselves trying to achieve conflicting objectives. For example, maximizing resource utilization may conflict with minimizing waiting times. In system optimization, one must be careful to weigh priorities and make sure the right objective function is driving the decisions. If, for example, the goal is to maximize resource utilization (keep them busy), then one need only keep the work queue in front of the resource continually full. Unfortunately, this builds up work-in-process and results in high inventory carrying costs in manufacturing systems. For service systems, large queues result in long waiting times, hence dissatisfied customers. On the other extreme, one might feel that reducing inventories or waiting times is the overriding goal and, therefore, decide to employ more than an adequate number of resources so that work-in-process or customer waiting time is virtually eliminated. Naturally, excessive resources can become quite costly and may not justify the resultant savings due to reduction in queue sizes. It is generally conceded that a better strategy is to find the best trade-off or balance between the number of resources and waiting times so that total costs are minimized (see Figure 2.3).

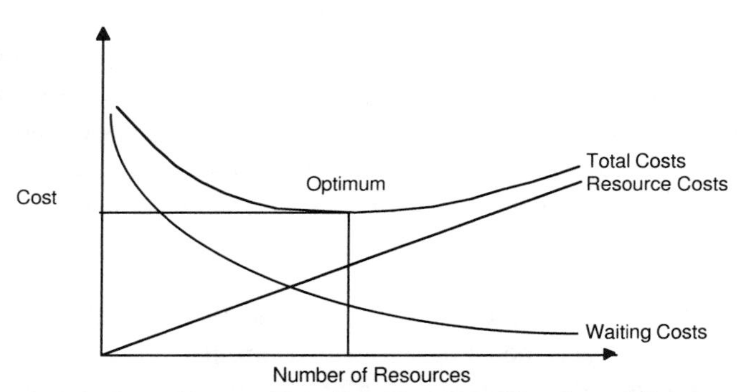

Figure 2.3 Cost Curves Showing Optimum Number of Resources to Minimize Total Costs.

As shown in Figure 2.3, the number of resources at which the sum of the resource costs and waiting costs is at a minimum is the optimum number of resources to have (it is also the optimum acceptable waiting time).

In systems design, the objective is not necessarily to arrive at the optimum system configuration. From a practical standpoint, the best that can be expected is to achieve a good working solution that reasonably accomplishes overall goals (see Figure 2.4).

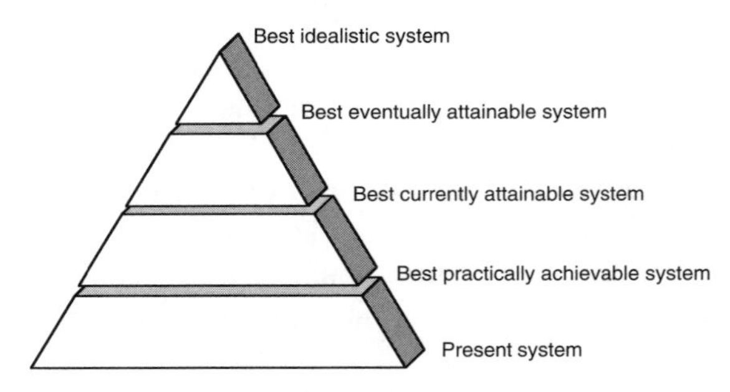

Figure 2.4 Systems Design Pyramid.
(adapted from Nadler 1965)

In the systems design pyramid shown in Figure 2.4, the *best idealistic system* is the perfect, but impossible, system having zero costs, instant response time, zero defects, absolute flexibility, and that which is completely self-controlled. The *best eventually attainable system* is one that is within the realm of future possibility, but is not currently practical given existing technologies. The *best currently attainable system* is the optimum system, or the best possible system that can be designed given current available technologies. While this would be desirable, it is usually not practical given the time and budget constraints as well as the decision making capability that are present during most design efforts. The *best practically achievable system* is the best system we can expect to be able to design given current constraints and decision making capabilities. In designing a system, we aim towards the best idealistic system, are guided by the best eventually attainable system, try to find the best currently attainable system, and expect to achieve the best practically achievable system (often we end up settling for something even less).

MODELS

The most common method used for supporting the decision making capability during the design stage is modeling. From a systems perspective, a model may be defined as a simplified representation of system relationships. This might include cause-and-effect relationships, flow relationships, and spatial relationships. The purpose of modeling is to understand, predict, control, and ultimately improve system behavior.

The art of modeling has been carried out for centuries. One example that occurred in 1585 was the relocation of the 340-ton Vatican Obelisk from the back to the front of St. Peter's Cathedral. Fontana, the engineer commissioned to perform the undertaking, had to figure out how the monolith could be lowered, moved, and raised. He modeled the process using a lead obelisk about two feet tall with a wooden tower, ropes, and tackle made to scale. Before relocating the actual obelisk, Fontana modeled each operation to uncover and find solutions to potential problems that could arise.

It must be remembered that even the best model is only an approximation of the actual system. A model therefore is neither true nor false, but rather useful or not useful. A useful model is one that serves the purpose for which it is intended. If the desired answers cannot be deduced from the model, then it is not very useful. The best model is one that achieves the desired purpose for the least amount of cost. A good model may, therefore, be characterized as follows:

- Includes only those elements that directly bear on the problem being solved.
- Is valid (accurately represents the system).
- Provides results that are meaningful and readily understood.
- Is easily modified and expanded.
- Is quick and inexpensive to build.
- Is credible (results are convincing to the customer).
- Is reusable.

In designing manufacturing and service systems, the modeling process is typically one in which the modeler analyzes an actual or proposed system and develops a concept of how the new or existing system works. This concept is translated into a model that is used to evaluate the concept. Through iteratively modeling, conceptualizing and relating back to the actual or proposed system, a good workable solution is found. This iterative process is illustrated in Figure 2.5.

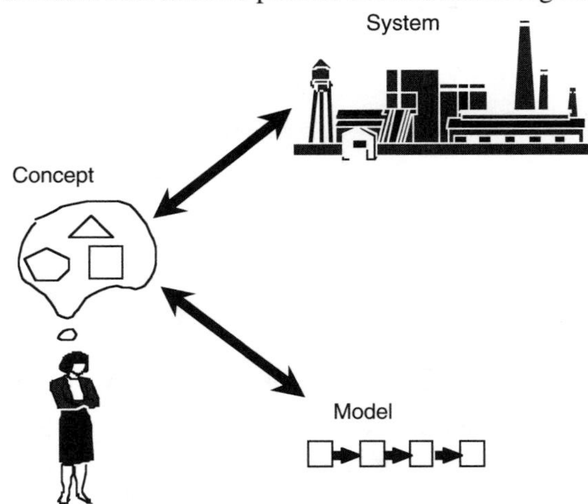

Figure 2.5 Iterative Nature of the Modeling Process.

TYPES OF MODEL

Models have been classified in many ways. Since we are dealing with dynamic systems modeling, it is useful to think of models as being one of the following three types:

1. Symbolic models
2. Analytic models
3. Simulation models

Each of these model types, along with their advantages and weaknesses, is described below.

Symbolic models

A symbolic model consists of graphic symbols such as rectangles and arrows used to depict activity sequences and other relationships. Examples of symbolic models are process flow diagrams, facility layouts, and IDEF0 diagrams. Symbolic modeling is perhaps the most popular technique for documenting processes in systems design and process reengineering. The reasons for the popularity of symbolic models are that they are quick and easy to develop and are easily understood by others. These static models are useful for general concepting and documentation but provide little or no quantitative analysis and frequently lack any operational detail.

Flow diagrams are a common way of modeling systems symbolically. One widely circulated procedure, defined by the American Society of Mechanical Engineers (ASME), utilizes the symbols shown in Table 2.1:

Table 2.1 Standard Process Flow Symbols.

Symbol	Description
○	Operation
⇨	Transportation
▽	Storage
D	Delay
□	Inspection
▢	Combination Operation and Inspection

Obviously, since symbolic models are used primarily for communication and documentation, there are no strict rules that must be followed as long as the desired information is communicated clearly and succinctly. The most simple diagrams consist only of boxes representing the activities performed and arrows depicting the activity sequence (see Figure 2.6).

Flow diagrams are extremely versatile and can be enhanced to include additional information such as resource requirements and activity times.

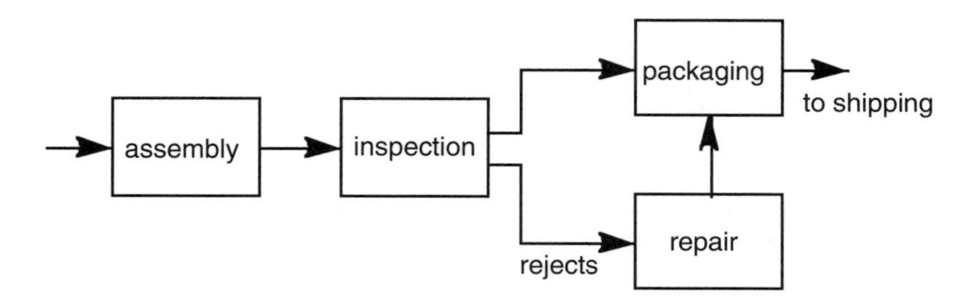

Figure 2.6 Example of a Flow Diagram.

One important benefit of symbolic models is that they help focus attention on the processes in a system without trying too early to resolve resource issues and operational problems. It does little good to apply new methods and technologies to outdated practices. Simply paving the cow path does not give us the best possible road. Shingo (1992) emphasizes the importance of focusing on process functions before detailed operation functions: "Improvements in process functions have a profound impact on operational functions. Some operations may be eliminated and some may be radically changed."

Shingo notes that due to the focus of Fredrick Taylor and Frank and Lillian Gilbreth on detail operational improvements, industrial engineering has become over preoccupied with operational analysis to improve production rather than focusing on process improvement (Shingo 1992). Process reengineering entails challenging all assumptions, questioning every step and pushing towards magnitudes of improvement. The major disadvantages of symbolic modeling are:

- They tend to lack detail.
- They provide little or no quantitative measure of system performance but merely describe elements, activities and relationships.
- They fail to capture the system dynamics.

Analytic Models

An analytic model is a mathematical formula that yields a quantitative solution. Analytic modeling can be simple arithmetic calculations on the back of an envelope or complex linear programming algorithms that provide optimum solutions for given sets of problems. Figure 2.7 is an example of an analytic model for calculating the expected number (*N*) of entities in a system consisting of a single server and single input queue into which entities arrive with a Poisson arrival rate (λ). Service times are assumed to be exponentially distributed with mean (μ):

$$N = \frac{\lambda}{\mu - \lambda}$$

Figure 2.7 Analytic Model for Calculating the Expected Number of Entities (N) in a Single Server, Single Queue System With a Poisson Arrival Rate (λ) and Exponential Service Time With Mean (μ).

Analytic models provide quick answers and some are able to give optimum solutions without going through trial-and-error. The major drawbacks to analytic models are:

- They often require that assumptions be simplified in order to fit the model.
- They are often unable to account for the random behavior that is exhibited in most systems.
- As systems become even moderately complex, the solutions get very complicated.
- They are incapable of solving problems of any great complexity.

Analytic models are generally static, prescriptive, and either deterministic or probabilistic. A linear programming model, for example, takes a given objective function and a set of constraints and calculates the best values for the decision variables. For a spreadsheet model, given a system requirement and resource capacities, the spreadsheet formula will derive an estimate of the number of resources needed. Other analytic models include Markov chain models, queuing models, and Petri net models. Many books have been written describing analytic solutions to systems problems such as Buzacott and Shanthikumar (1993), Gershwin (1994), and Askin and Standridge (1993).

Several products have been developed to provide analytic solutions to system problems. Rapid Modeling Technology (RMT) is an example of an analytic modeling technique which is deterministic and provides "rough cut" information. RMT takes information on routings, machines, mean time between failure (MTBF) and mean time to repair (MTTR) (averages only) and estimates of the scrap and rework for each family. Based on this information, the method predicts machine utilizations, product lead times and work-in-process. A product similar to RMT is Manuplan which provides an estimate of the average time required for a product to move through a factory (see Suri and Tomsicek 1988).

Simulation Models

Simulation is a modeling technique in which the cause-and-effect relationships of a system are captured in a computer model, which then becomes capable of generating the same behavior that would occur in the actual system. The simulation produces an actual history log and statistical summary of all activity that took place in the model over a designated period of time. The output results from running the simulation give quantitative measures of system performance, such as resource utilization and cycle times. In this sense, simulation is an evaluation tool.

A simulation model should be viewed essentially as a "what if" tool that allows a designer or manager to experiment with alternative designs and operating strategies to see what impact those decisions have on overall system performance. As an experimental tool, simulation is used to test the effectiveness of a particular design and does not, in itself, solve a problem or optimize a design. It helps evaluate a solution and provides insight into problem areas rather than generate a solution. Arriving at an optimum (or rather near optimum) solution can only be obtained through experimentation by running and comparing the results of alternative solutions.

A simulation model, using today's simulation tools with animation and graphical output, not only provides a symbolic or graphical representation of a system, but also graphically shows the dynamic behavior of the system over time (see Figure 2.8). Like analytic modeling, simulation is quantitative and provides measures on resource utilization and waiting times. Unlike symbolic or analytic modeling, simulation models are able to take into account the most complex statistical fluctuations and interdependencies. Simulation helps eliminate the subtle inefficiencies and hidden waste that are not obvious from a diagram and often go undetected until the system is in operation and it is too late.

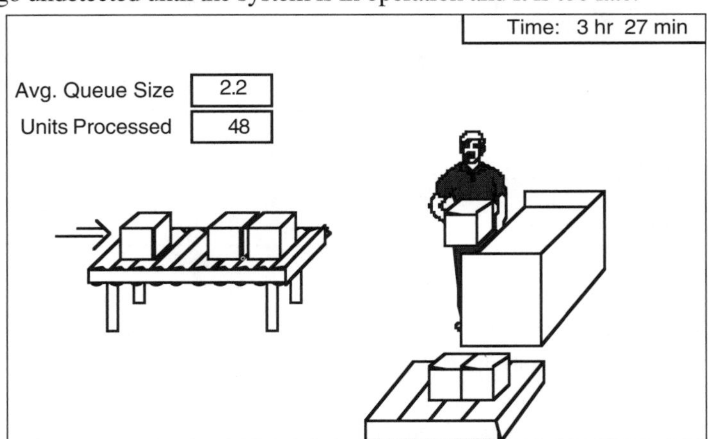

Figure 2.8 Example of a Simulation Model with Animation.

Computer or digital simulation works by executing a computer program that mimics the behavior of the real system. Analytic methods, on the other hand, attempt to formalize the main dynamics of the system into an equation or heuristic. To the extent that symbolic and analytic modeling have been integrated into simulation software, simulation software can provide many of the same benefits as symbolic and analytic modeling, in addition to its own unique modeling characteristics. The real power of simulation lies in its ability to measure the effects of variation and interdependencies on overall system performance. The primary disadvantages of simulation models are:

- They can be difficult to construct.
- They are more descriptive than prescriptive.

SELECTING THE APPROPRIATE MODEL

There is no single "best" modeling technique to use, but rather each modeling method has strengths that lend it to particular situations or phases in systems analysis and design. The appropriate model for a given problem should be selected by assessing the objectives in light of the capabilities and limitations of each modeling methodology. Table 2.2 compares the characteristics of each modeling method.

Table 2.2 Comparison of Symbolic, Analytic, and Simulation Modeling.

	Symbolic	Analytic	Simulation
Prescriptive	no	yes	somewhat
Quantitative	no	yes	yes
Dynamic	no	no	yes
Easy to use	yes	somewhat	somewhat
Flexible	yes	no	yes
Models Randomness	no	somewhat	yes
Models Inter-dependencies	no	somewhat	yes

● yes ○ no ⊘ somewhat

A designer need not necessarily choose one modeling method at the exclusion of the other two. On the contrary, these methods work well in concert and actually complement one another. For example, a system design may begin with a symbolic model to help conceptualize and design the process. From this, analytic models (when available) may be used to arrive at "rough cut" calculations of system requirements. Finally, a simulation model might be constructed to verify the decisions, fine-tune the design, and look for hidden opportunities for improvement. Figure 2.9 illustrates this successive modeling technique.

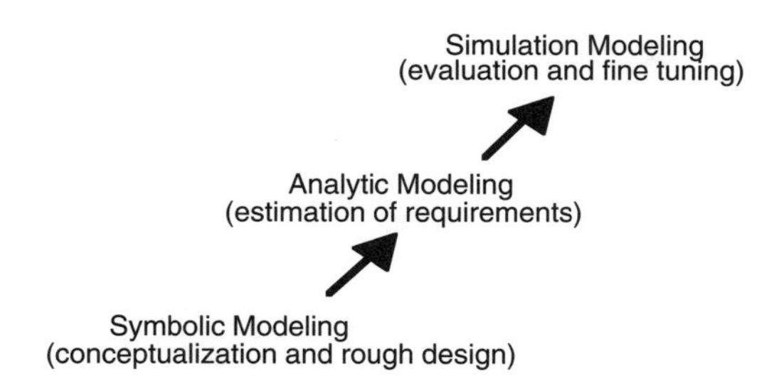

Figure 2.9 Successive Use of Symbolic, Analytic, and Simulation Modeling.

Often the use of a second model may be helpful in establishing the validity of the first model. This cross checking procedure can at least verify whether the results of a model are in the ballpark.

SUMMARY

In this chapter, systems and models were examined and the use of modeling in analyzing and designing systems was discussed. The importance of taking a systems approach to designing systems was emphasized. The systems approach looks at all of the cause-and-effect relationships that have a bearing on overall objectives. An effective design is based on accurate prediction of performance in response to certain decision variables.

Systems are modeled using either symbolic, analytic, or simulation models. Simulation modeling is the most powerful method since it provides a reliable estimate of system performance and is capable of modeling virtually any system configuration or dynamic complexity. All three modeling methods can be beneficially used to help in systems planning.

Chapter 3
Simulation Basics

"Nothing is more terrible than activity without insight."

Thomas Carlyle

INTRODUCTION

Simulation is a relatively easy tool to use, but by the same token it can be quite complex — complex but not overly complicated. Simulation must be able to model complex situations and decision making processes. This chapter presents a basic overview of how simulation works. This foundation is essential to understanding what simulation is actually doing. Each simulation language has certain subtleties that affect the way entities are processed and events take place. Output results are also somewhat general and must sometimes be customized to get the desired measures. For these reasons, the user must have at least a fundamental understanding of how simulation works. A manual simulation example is presented at the end of the chapter to illustrate what happens in a simulation. For additional reading about simulation, there are many excellent books that are available (see Banks and Carson 1984; Gottfried 1984; Hoover and Perry 1989; Law and Kelton 1991; Ross 1990; Thesen and Travis 1992; Shannon 1975; and Widman, Loparo and Nielsen 1989).

WHAT IS SIMULATION?

Simulation in the broad sense has been defined as "an activity whereby one can draw conclusions about the behavior of a given system by studying the behavior of a corresponding model whose cause-and-effect relationships are the same as (or similar to) those of the original" (Gottfried 1984). Simulation uses a computer program to actually mimic causal events and the consequent actions in a system. Statistics are accumulated during the simulation on measures of interest which are summarized and reported at the end of the simulation. As described by Banks and Carson (1984): "Simulation involves the generation of an artificial history of a system, and the observation of that artificial history to draw inferences concerning the operating characteristics of the real system."

Those unfamiliar with simulation often have the erroneous notion that simulation itself solves problems. They assume that by defining the problem, simulation will generate a solution. Such techniques are found in *expert systems* and generally consist of rules and heuristics defined by an expert for providing solutions to given sets of problems. Expert systems are essentially "black box" approaches to problem solving because the user does not need to know — and is often not even interested in knowing — how the solution is derived.

Simulation is a solution evaluator, not a solution generator. It does not produce an optimum theoretical solution but rather directs one towards the best workable solution. In the literature on artificial intelligence, this approach of building and analyzing computational models is referred to as model-based reasoning. It differs from expert systems in that the user must iteratively define, simulate, and analyze until a solution is reached.

To a certain extent, expertise can be built into a simulation language in the form of heuristics and optimizing algorithms that perform some of the decision making that would normally be performed by an expert. Many simulation languages, for example, are able to automatically determine the shortest route in a path network for entities traveling between two points. Where such expertise is built into the language, the user is relieved from having to understand just how such decision logic is implemented. Because of the complex nature of simulation and the multiple objectives one has when using simulation, it is unlikely that simulation will ever fully replace an expert except for narrow classes of applications.

DISCRETE-EVENT VERSUS CONTINUOUS SIMULATION

Simulation software has traditionally been classified as being either discrete-event or continuous. *Discrete-event* simulation products are used for modeling systems that change state at discrete points in time as the result of specific events (see Figure 3.1). Most manufacturing and service systems are discrete-event systems.

Continuous simulation software is used to model systems whose state changes continuously with respect to time, such as the temperature of an ingot in a furnace or the volume of a tank in a beverage production facility. Continuous simulation products use differential equations to calculate the change in a state variable over time. These equations are usually solved at small time-step increments to determine the current values of state variables until some threshold is reached which initiates some action.

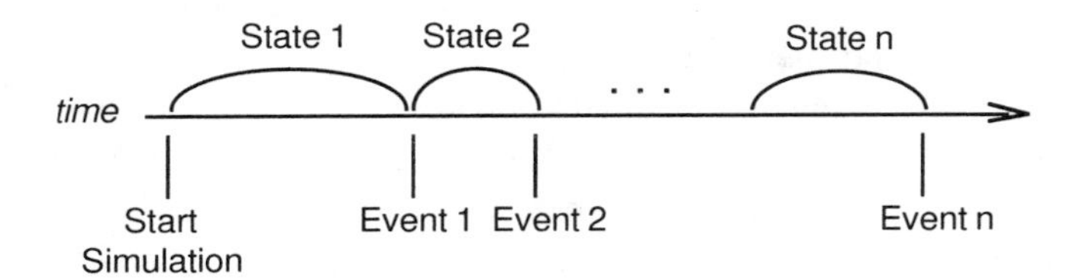

Figure 3.1 Progression During Simulation From Event to Event.

Some simulation software products have both discrete-event and continuous modeling capabilities. This allows systems that have both discrete-event and continuous characteristics to be modeled.

When speaking of the state of the system, we are not talking about whether the system as a whole is busy, down, or idle. The state of a system is actually described by the values of all of the individual state variables within the system. A state variable might be the number of entities in a particular queue or the status (busy, idle, down) of a particular resource. State variables in a discrete-event simulation are referred to as *discrete-change* state variables. Continuous systems have *continuous-change* state variables that change continuously over time, such as the temperature of a building being controlled by a heating and cooling system.

Often it is possible to model continuous phenomena using discrete-event logic, especially if a high degree of precision is not important. For example, continuous flowing substances such as liquids or granules can be converted, for purposes of simulation, into discrete units of measure such as gallons or pounds. Another method is to simply update a variable at regular time intervals that accounts for a constant rate of change that occurred over the interval.

Just as continuous state changes can be modeled using discrete-event logic, it is possible, and sometimes even desirable to treat discrete-change states that occur at very small intervals (such as a high-rate processing system) using continuous-change state logic.

Throughout this book, simulation will be used synonymously with discrete-event simulation as opposed to continuous simulation.

STOCHASTIC VERSUS DETERMINISTIC SIMULATION

As pointed out in the first chapter, one of the most powerful features of simulation is its ability to model random behavior or variation (operation times, reject rates, arrival intervals).

Nearly every manufacturing or service system exhibits some type of randomness. Table 3.1 identifies variables in a system that are often random in nature.

Models that are based on one or more variables that are random in nature are referred to as *stochastic* models. A stochastic model produces output that is itself random and therefore only an estimate of the true behavior of the model. Models that have no input components that are random are said to be *deterministic*. A deterministic model is one in which the behavior of the model is "determined" once the input data has been defined. Actions in a deterministic model are always the same and always result in the same outcome. Deterministic simulation models are built exactly the same as stochastic models except that they contain no randomness. The primary difference lies in the method of analyzing the results. In a deterministic model, the result of a single simulation run is an exact measure of the performance of the model. For stochastic models, several sample runs must be made, and then the composite average result provides only an estimate of the expected performance of the model.

Table 3.1 Typical Random Variables in a System.

Random Variable	Examples
Activity durations	Operation times, repair times, setup times, move times
Activity or decision outcomes	The success of an operation, the decision of which activity to do next
Quantities	Lot sizes, arrival quantities, number of workers absent
Event intervals	Time between arrivals, time between equipment failures
Attributes	Customer type, part size, skill level

To say that a model variable is random does not mean that it is undefinable nor even unpredictable, but rather the phenomenon being modeled tends to vary statistically. Statistically varying phenomena are probabilistically predictable. Describing a random phenomenon in a model is done by specifying either a probability expression or a probability distribution depending on whether the random phenomenon is the likelihood of an outcome (e.g., pass or fail) or the likelihood of a value (e.g., activity time or batch size).

SIMULATING PROBABILISTIC OUTCOMES

For decision outcomes that are probabilistic, such as when a decision results in a particular choice or outcome a certain percentage of the time, each time the decision is made during the simulation, a sample value is randomly drawn from a continuous, uniform (any value is equally likely to occur) distribution between 0 and 1 ($0 \leq X < 1$). The formula for generating a random number between 0 and 1 is called a *random number generator*. The value of the sample drawn determines the outcome of the decision. To illustrate how a probabilistic outcome is determined, suppose that parts pass an inspection operation 80 percent of the time and fail the inspection 20 percent of the time (see Figure 3.2).

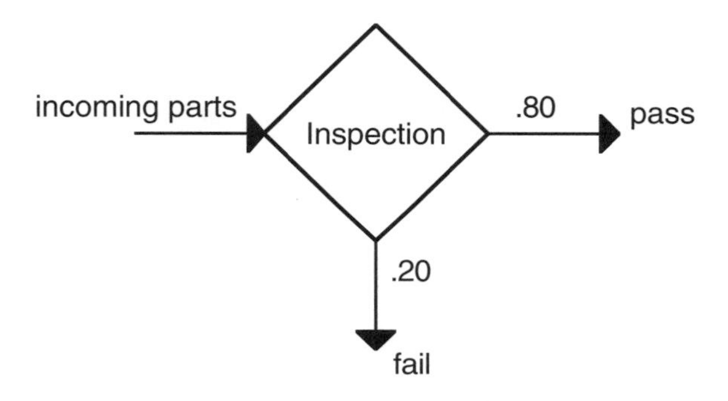

Figure 3.2 Diagram Showing Probability of Passing or Failing an Inspection Operation.

In the model, the probability of each outcome is defined as being .80 and .20. During the simulation, whenever that particular inspection occurs, a random number is generated between 0 and 1. If the number generated is between 0 and .8 (say .52), the outcome of the inspection decision is to pass the part; otherwise the outcome of the decision is to fail the part.

Simulating Statistically Varying Values

Stochastic systems frequently have time or quantity values that vary statistically from occurrence to occurrence. Probability distributions express both a probability, or shape, and a range of values. They are useful for predicting the next time, distance, quantity, etc. to use for a particular random variable in the simulation, such as the mean time between failures for a machine. Probability distributions differ from simple probability expressions in that they are expressed with more than one parameter in order to define the shape and range of the distribution. For example, we might describe the time for a check-in operation to be normally distributed with a mean of 5.2 minutes and a standard deviation of .4 minutes. During simulation a sample is drawn from this distribution whenever the next operation time must be determined. The shape and range of time values generated for this activity will correspond to the parameters used to define the distribution. Values drawn from a particular distribution are referred to as *random variates.* Most of the methods used in simulation products for generating random variates can be found in Law and Kelton (1991).

Probability distributions may be either discrete (they describe a range and likelihood of possible discrete values) or continuous (they describe a range and likelihood for a continuous range of possible values). Figure 3.3 illustrates a graph for both discrete and a continuous distribution.

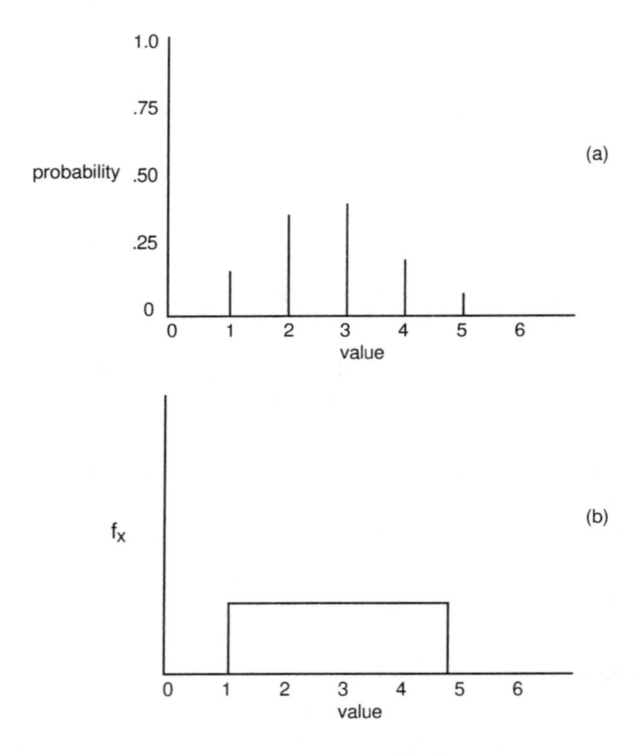

Figure 3.3 Example of a Discrete Probability Distribution *(a)* and a Continuous Probability Distribution *(b)*.

An example of the use of a probability distribution is a machine with a time between failures that is exponentially distributed with a mean of 30 minutes. In real life we do not know exactly when it is going to go down until it actually happens. However, since we know the statistical distribution of the time between failures, we can pick a random variate from the distribution to determine the next downtime occurrence.

HISTORY OF SIMULATION

Computer simulation was first put to use in the aerospace industry in the 1950s. In the 1960s, it began to be applied to industrial systems although models were sometimes crude. There are a variety of reasons for this slow start including the following:

- Computers were expensive.
- Modeling required extensive programming.
- Processing speeds were slow.
- Memory was constrained.

Early simulation models were developed primarily using general programming languages such as FORTRAN.

As interest in simulation increased, it became obvious that many systems shared similar characteristics that could be incorporated into a general modeling language. Several simulation languages were developed (GPSS, SIMSCRIPT, SIMAN, SLAM, etc.) for modeling a class of systems referred to as *queuing systems.* These languages provided features for defining basic entity processing times, resource usage and queuing situations. General reporting capabilities were also provided.

As simulation spread to different industries during the 1980s, more industry-specific simulation products (AutoMod, ProModel, WITNESS) appeared that incorporated the characteristics of an industry sector. Because many of these emerging products were data-driven and lacked programming capability, they were referred to as *simulators* rather than languages. Some of these data-driven products, however, now provide full programming capability, thus combining the flexibility of a simulation language with the ease-of-use of a simulator.

Simulation products are continuing to become more adaptable and easier to use as some products are now oriented towards product- or service-specific industries, such as semiconductor manufacturing or fast food services. Models can be built with such products in a matter of a few hours using modular constructs that have already been developed and pretested. Because they use terminology specific to a particular industry, it is much easier for people doing simulation in those industries to learn and use the products.

Simulation products are also expanding to facilitate more of the overall simulation effort including the following:

- Distribution fitting routines for input data
- Graphical input of model components
- Dialog boxes for guiding input

- Online help and tutorials
- Interactive debugging
- Logic and completeness checking
- Automated model development
- Goal-driven experimental design modules
- Full output statistical analysis and output charts

These features provide a customizable, completely integrated environment that makes simulation much more of a system rather than merely a language.

Another trend in simulation has been *visual interactive simulation* in which the user of the model actively participates in the simulation. Decisions can be made while the simulation is running regarding staffing levels, work assignments, etc. This "human in the loop" modeling procedure is useful for doing rough experimentation and concepting. Obviously, statistical output from such simulations is relatively meaningless since the basis for gathering statistics changes during the run. If, for example, the capacity of a queue is altered during the run, the utilization statistic for the queue would not have a single capacity value upon which it could be based. Figure 3.4 illustrates the evolution of simulation languages over the past several decades.

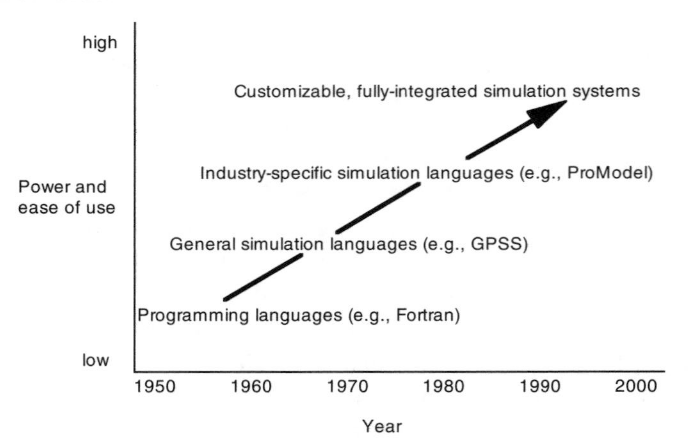

Figure 3.4 Evolution of Simulation Tools.

COMPONENTS OF SIMULATION SOFTWARE

Simulation software consists of several modules for performing different functions during simulation modeling (Figure 3.5). Typical modules found in simulation software are described as follows:

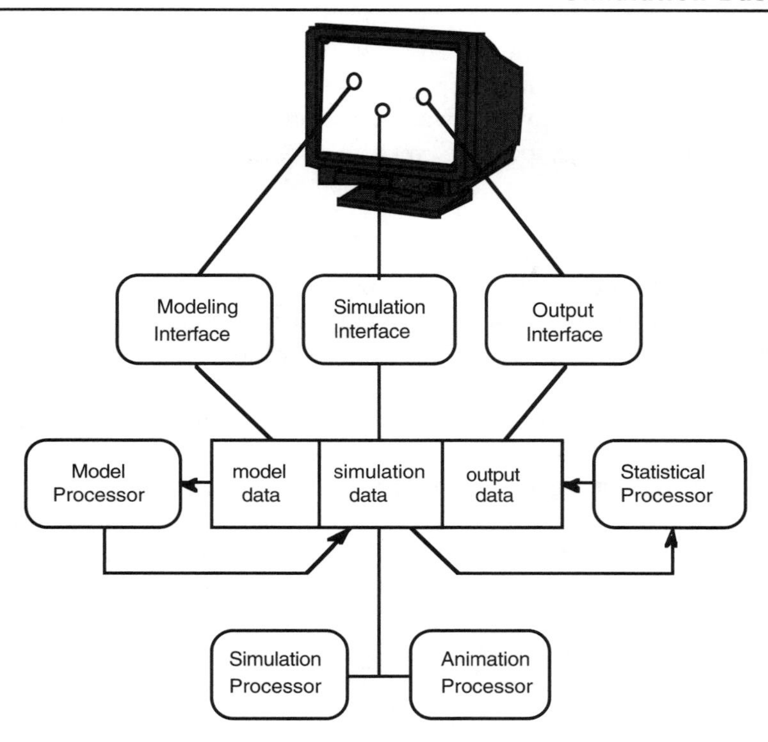

Figure 3.5 Typical Components of a Simulation System.

Modeling Interface Module

The user defines the model using the input or modeling interface module. This module provides graphical tools, dialogs and other text editing capabilities for entering model information. The model database is generally in a form that is readable and easily edited.

The language features provided for defining particular system elements are referred to as the language's *modeling constructs*. Modeling constructs may be thought of as the basic building blocks for defining models. Some simulation languages have more modeling constructs than others. Likewise, the type of constructs provided may be different from software to software. Some languages provide fairly sophisticated constructs that permit models to be defined at a relatively high level without requiring lots of detailed coding. This approach enables models to be built quickly and easily, but carries the risk that the constructs may be oversimplified or too restrictive for certain situations. Other languages have more simple constructs, but they are usually quite primitive and require an inordinate amount of low level piecing together to achieve the desired effects. These languages are flexible and permit special situations to be modeled.

However, model building is difficult and time consuming and often requires extensive debugging time. The most powerful simulation languages of today provide a wide range of constructs that permit high level definitions where possible, yet also provide low level capability for handling special modeling situations.

Model Processing Module

Once a model is defined, the model processing module takes the model database, and any other external data files that are used as input data, and creates a simulation database. This data conversion is performed because the model database and external files are generally in a format that cannot be used very efficiently by the simulation processor during simulation time. In addition to the translated model input data, other data elements are created for use during simulation including statistical counters and state variables. *Statistical counters* are used to log statistical information during the simulation such as the cumulative number of entries into the system or the cumulative activity time at a workstation. *State variables* collectively describe the state of the system at any given instant in time. Examples of state variables include the following:

- Current entities waiting in a queue
- Current state of a machine (busy, idle, down)
- Current number of resources in use
- Current number of entities in the system

State variables often have an effect on the decisions that are made when events occur. For example, customers might choose which checkout lane to enter based on the number of customers in each lane. A change in a state variable of the system may also cause other events to occur. When an inventory level drops below a certain level, for example, a replenishment order is issued.

Simulation Interface Module

The simulation interface module displays the animation that occurs during the simulation run. It also permits the user to interact with the simulation to control the current animation speed, trace events, debug logic, query state variables, request snapshot reports, or pan or zoom the layout. If visual-interactive capability is provided, the user is even permitted to make changes dynamically to model variables with immediate visual feed-back of the effects of the changes on the simulation.

Simulation Processor

The simulation processor processes the simulated events and updates the statistical counters and state variables. A typical simulation processor might consist of the following components:

- *Clock variable*—a variable used to keep track of the elapsed time during the simulation.
- *Scheduled events list*—a list of scheduled events arranged chronologically according to the time they are to occur.
- *Conditional events lists*—one or more lists containing events waiting for a state condition to be satisfied before continuing processing.
- *Event dispatch manager*—the internal logic used to update the clock and manage the execution of events as they occur.
- *Event processing algorithms*—algorithms that describe the logic to be executed when an event occurs.
- *Random number generator*—an algorithm for generating one or more streams of pseudo random numbers between 0 and 1.
- *Random variate generators*—routines for generating random variates from specified probability distributions.

Animation Processor

The animation processor updates and displays the graphical representation of the activity taking place during the simulation. The animation is often displayed during the simulation itself, although some simulation products create an animation file that can be played back without the simulation. Some simulation tools display history plots that are dynamically updated during the simulation.

Animation and dynamically updated displays and graphs provide a visual representation of what is happening in the model while the simulation is running. Animation comes in varying degrees of realism from three-dimensional animation to simple animated flow charts. Often, the only output from the simulation that is of interest is what is displayed in the animation. This is particularly true when simulation is used for basic concepting or for communication purposes.

For most simulations where statistical analysis is required, animation is no substitute for the post-simulation summary that gives a quantitative overview of the entire system performance.

Output Processor

The output processor summarizes the statistical data collected during the simulation run and creates an output database. Typical kinds of data summarized include:

- Resource utilization
- Average queue sizes
- Average operation times
- Average time in system
- Number of entities processed per period

Most simulation software also provides the ability to write data out to external files completely defined by the user.

Output Interface Module

The output interface module provides a user interface for displaying the output results from the simulation. Output results may be displayed in the form of reports, plots, histograms, or pie charts. Output data analysis capability such as correlation analysis and confidence interval calculations is also often provided. Some simulation products even point out potential problem areas such as bottlenecks.

HOW DISCRETE-EVENT SIMULATION WORKS

In order to understand how simulation works, it is important to understand the concept of simulation events. Discrete-event simulation works essentially by translating the model data logic into cause-and-effect relationships, and then determining, one at a time, when the next simulation event occurs that causes some logic to be processed and other events to happen. Simulation events may be triggered by a condition, an elapsed time or by some other chain of events (see Figure 3.6). Typical simulation events might include:

- Arrival of an entity to a workstation
- Failure of a resource
- Completion of an activity
- End of a shift

Simulation events are of two types:

1. *Scheduled events* that occur at times that are scheduled (a process completion or a machine failure).
2. *Conditional events* that occur only when one or more conditions have been satisfied or events have occurred (the departure of a customer from a waiting line when a service agent becomes available, or the shipment of an order when all of the line items have been pulled and consolidated).

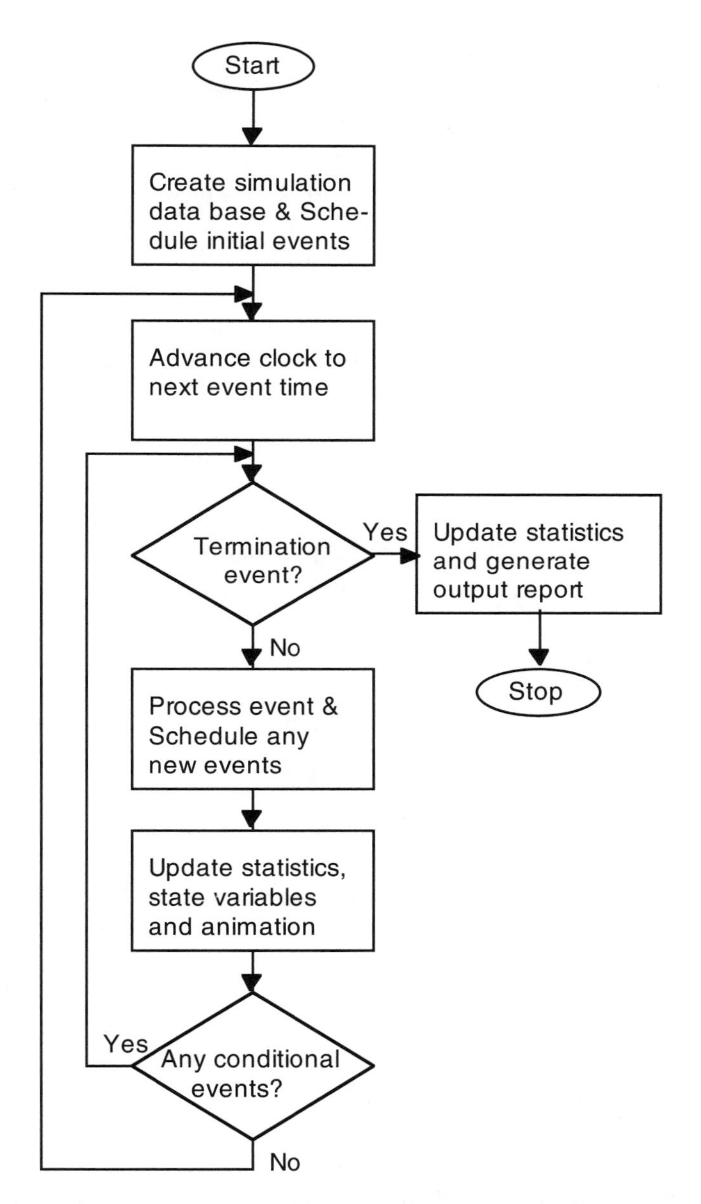

Figure 3.6 Logic Diagram of How Discrete-event Simulation Works.

Scheduled event times are typically determined by randomly sampling from an input probability distribution that describes the activity time or time between occurrences for each particular type of event. These events are usually scheduled at the moment they can be predicted. For example, as soon as an activity begins that takes a specified amount of time, a completion event can be scheduled.

Simulation works essentially by arranging the initially scheduled events chronologically into the scheduled events list. The clock is then updated to the first event in the list and the logic associated with that event gets processed. The processing of an event, whether it be a scheduled event or a conditional event, consists of updating the affected state variables in the system, collecting associated statistics and, if animation is included, updating the screen picture. Any consequent, scheduled events are placed into the scheduled events list. Likewise, there may be consequent conditional events resulting from the processing of the event that are put into appropriate waiting lists.

After a scheduled event is processed, any conditional event whose condition is now satisfied also gets processed. When no more conditional events are able to be processed, the clock advances to the next scheduled event time. When an end of simulation event (which might have been defined as either a scheduled or conditional event) occurs, the simulation terminates and statistical reports are generated. The updating of the clock to the next imminent event, the processing of logic and state changes associated with each event, and the collection of statistical data constitute the essence of discrete-event simulation.

SIMULATION EXAMPLE

To illustrate how simulation works, we will now present a manual simulation of a simple system in which computer monitors get packaged prior to going to shipping. The packaging station has an input queue which, for all practical purposes, has unlimited capacity. The package station itself has a capacity of one (only one monitor at a time can be packaged). Monitors are placed into the input queue as they arrive from manufacturing with an interarrival time that is exponentially distributed (they occur completely randomly and independently of each other) with an average time between arrivals of 1.2 minutes. If the package station is empty or available, a monitor is pulled from the queue and packaged. If the package station is busy and therefore unavailable, the monitor waits in the input queue. Packaging is automated and takes one minute, after which the monitor is sent to shipping. From a simulation standpoint, we can simply say that they "exit" the system. A material flow diagram is shown in Figure 3.7.

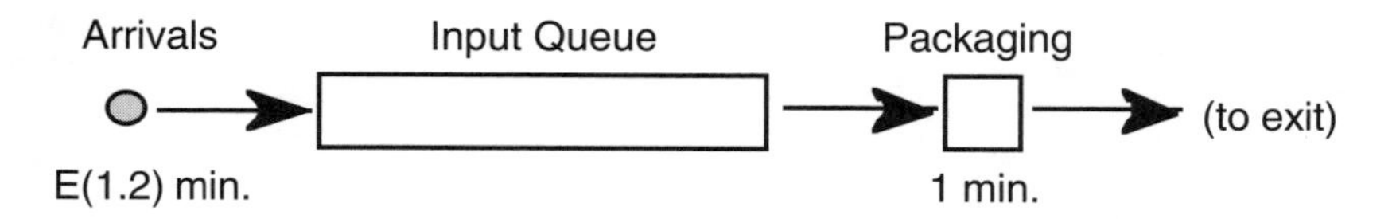

Figure 3.7 Entity Flow Diagram for Example System.

Model Assumptions

As in any simulation, certain assumptions must be made where information is unclear or incomplete. Note the assumptions we will be making for this simulation:

1. There are no parts in the system initially so the queue is empty and the packaging station is idle.
2. Activity is never constrained by personnel; no operators are included in the model.
3. The move time from the queue to the package station is negligible and therefore ignored.
4. Monitors are pulled from the queue on a first-in first-out (FIFO) basis.
5. The package station never experiences failures.

Running the Simulation

The activity will be manually simulated for ten minutes with the objective of finding the average time parts spend in the queue. Recall that discrete-event simulation processes the events as they occur over time. As these events get processed, the state variables (number in the queue and package station status) get changed and the statistical accumulator for the cumulative time in the queue gets updated. A simulation logic diagram for this example is shown in Figure 3.8.

In Figure 3.8, the first thing that is done is to schedule a termination event to occur after 10 minutes and an arrival event to occur at 1.1 minutes (arrival times are assumed to be generated from a table of random values that are exponentially distributed with a mean of 1.2 minutes). There are two types of recurring scheduled events: *arrival events* and *packaging completion events.*

The processing of an arrival event initiates the creation and arrival of a monitor at the input queue. Upon entering the queue, the contents of the queue are incremented and the cumulative queue time updated (since this is the first entry, there is no cumulative time yet). The packaging station is available so the monitor immediately exits the input queue which again updates the queue contents and cumulative queue time (these values remain zero since all of this happened at the same instant in time).

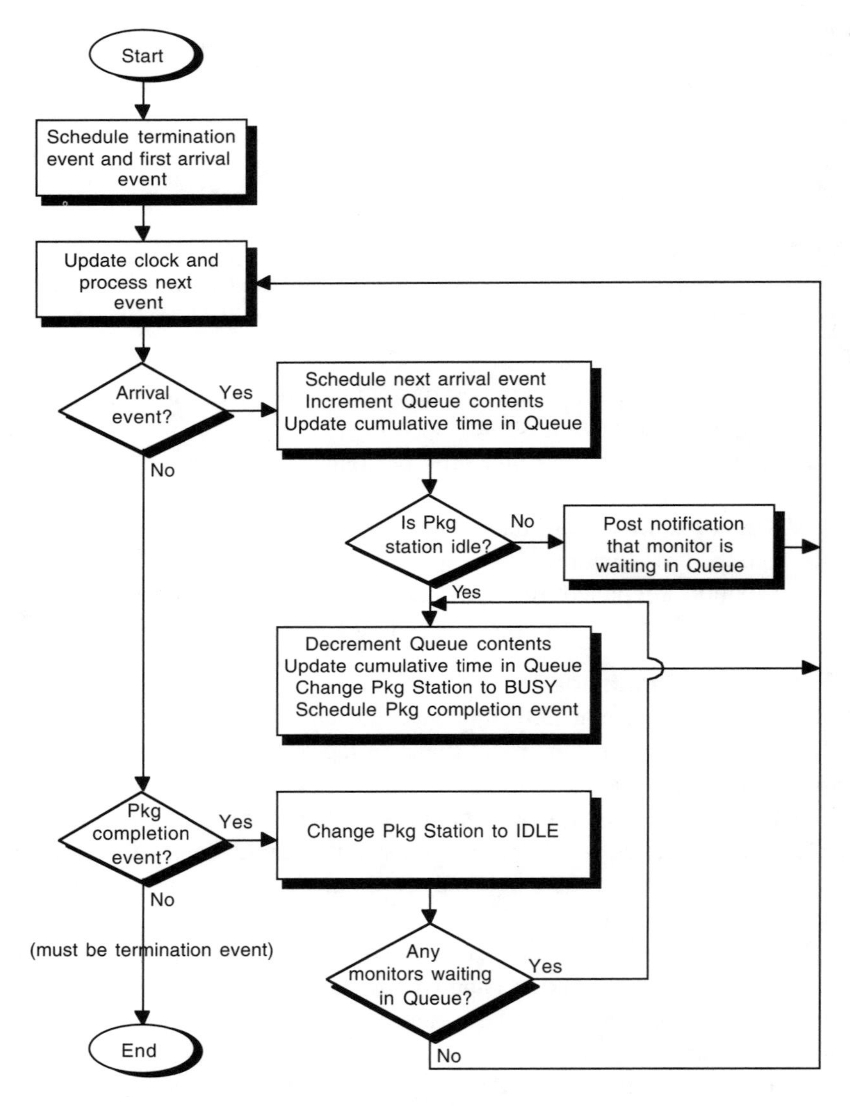

Figure 3.8 Simulation Logic Diagram for Example Model.

Upon entering the packaging station, the station status is set to *busy* and a packaging completion event is scheduled. The monitor waits at the packaging station until the packaging completion event is processed. Meanwhile, the clock is updated to the time of the next event (2.1 minutes) and the next event (a packaging completed event) gets processed. A history of the simulation activity is recorded in Table 3.2.

Table 3.2 Simulation History of Manual Simulation.

Event time (minutes)	Unit number	Event type	Number in queue	Cumulative time (min) in queue	Pkg station status
1.1	1	arrival	0	0	busy
2.1	1	pkg done	0	0	idle
2.8	2	arrival	0	0	busy
3.5	3	arrival	1	0	busy
3.8	2	pkg done	0	.3	busy
4.1	4	arrival	1	.3	busy
4.6	5	arrival	2	.8	busy
4.8	3	pkg done	1	1.2	busy
5.6	6	arrival	2	2.0	busy
5.8	4	pkg done	1	2.4	busy
6.8	5	pkg done	0	3.4	busy
7.4	7	arrival	1	3.4	busy
7.8	6	pkg done	0	3.8	busy
7.8	7	pkg done	0	3.8	idle
9.1	8	arrive	0	3.8	busy
9.8	9	arrive	1	3.8	busy
10.0	-	terminate	1	4.0	busy

The *unit number* in Table 3.2 is the particular monitor being processed at this event time. The *number in the queue* reflects the current quantity of monitors in the queue at the end of processing the event. The *cumulative time in the queue* is calculated by multiplying the contents of the queue at the end of the previous event by the elapsed time since the previous event. This value is then added to the cumulative time at the end of the previous event.

This simple and brief simulation is relatively easy to follow. Imagine a system with dozens of additional processes with many more factors influencing behavior such as downtimes, preemptions, resource contention, etc. You can see how computers really made simulation a practical analysis tool. Computers have no difficulty tracking the many relationships and updating the numerous statistics that are present in most simulations. Equally as important, computers are not error prone and can perform millions of instructions per second with absolute accuracy.

Calculating Results

When the simulation is completed, the statistics for the system behavior can be calculated. Note that there were a total of 9 arrivals over the simulation period. Of the 9 monitors that entered the system, only 6 monitors actually waited in the queue. Also note that there are still 2 monitors in the system at the end of the simulation. The average waiting time for monitors that spent any time in the queue is .67 minutes (4 minutes/6 parts). The average waiting time for *all* monitors that entered the system is .44 minutes (4 minutes/9 parts).

Like the statistics calculated for average waiting time, other statistics can be determined, such as average utilization of the packaging station. This can be calculated by adding up all the time intervals during which the packaging station was busy (6.9 minutes) and dividing this total busy time by the total simulation time (10 minutes). The resultant utilization is .69 or 69 percent.

Observations

Even though this is a simple simulation run, it provides an excellent illustration of basic simulation issues that need to be addressed when conducting a simulation study. *First*, note that the simulation startup and ending conditions can bias the output statistics. Since the system started out empty, queue content statistics are slightly less than what they might be if we began the simulation with parts already in the system. Likewise, the system ends with one part still in the queue so that if we base statistics on total queue time, part of this time reflects only a partial time in the case of the last entity. *Second*, note that we ran the simulation for only ten minutes before calculating the results. Had the simulation run longer, it is very likely that the long run average time in the queue would have been somewhat different than the short run, since the short run simulation did not have a chance to reach statistical regularity.

These are the kinds of observations that should be made whenever any simulation is run. The modeler must carefully analyze the output and ask questions about the meaning of the results.

SUMMARY

Discrete-event simulation works by converting all activity to events and consequent reactions. These events are processed one at a time until the simulation ends. Simulation software consists of several modules with which the user interfaces. Internally, model data is converted to simulation data which is processed during the simulation. At the end of the simulation, statistics are summarized in an output database that can be tabulated or graphed in various forms.

Chapter 4
Getting Started With Simulation

"When you fail to prepare, you prepare to fail."

Unknown

INTRODUCTION

Having shown the benefits of simulation and explained how simulation works, we are ready to look at how companies incorporate simulation into their business practices. The purpose of this chapter is to provide a guideline for getting started with simulation once the decision to adopt the technology has been made. The importance of management commitment is discussed along with the other steps necessary to successfully incorporate simulation into the decision-making activities of a company. Important features to look for in simulation software are identified with guidelines for properly evaluating and selecting a simulation product.

Businesses sometimes get into simulation haphazardly, with little direction. Prior to using simulation on any project, the proper tools, skills and resources must be in place. Many simulation efforts are doomed to failure from the outset because of poor planning and lack of appropriate tools. If adequate simulation tools are acquired, and properly trained personnel are put in place, the chances of achieving success in the use of simulation can almost be guaranteed.

MANAGEMENT COMMITMENT

As is true with any new technology, the most important ingredient for successfully incorporating simulation into a company's decision-making processes is management commitment. It takes time and patience to begin reaping the benefits of a new technology. Without the commitment and support of management, a technology such as simulation stands a only a slim chance of succeeding. This commitment has to be more than lip service and fleeting gestures. It must take the form of positive, sustained action by providing the necessary planning, resources, funding and follow-up to make it happen.

One of the best ways for management to become committed to simulation is to become informed both in the benefits of using simulation as well as in the process of doing simulation. This understanding will help prevent any misguided expectations that management might have with regard to the technology. Many simulation projects fail because management expects a three month project to be completed in three weeks. Other failures occur because management does not understand the need for qualified people to be involved in simulation and, hence, utilizes untrained staff members who must often do simulation projects as second priority to their normal responsibilities.

STEPS FOR GETTING STARTED

Once management has a basic understanding of the technology of simulation and recognizes the need for adopting simulation within the business, the necessary steps can be taken for getting started in the use of simulation. While there is no set procedure for getting started to do simulation, we have found the following steps to be beneficial:

Step 1	Appoint a simulation champion.
Step 2	Determine simulation needs.
Step 3	Assess current simulation technologies.
Step 4	Evaluate and select simulation software.
Step 5	Acquire hardware and any additional software.
Step 6	Conduct a pilot project.

Step 1: Appoint a Simulation Champion

In every business where simulation has been successful there has been at least one simulation "champion" involved who had the vision of what simulation could do and who internally promoted the technology. This should be an individual whose enthusiasm is contagious—one who not only has the aptitude for understanding the technology and how to apply it, but also is capable of winning over the skeptics. The enthusiasm and vision of the champion is crucial to ensuring that the conservative, resistant forces in a company do not hinder or prevent the acceptance of the technology. The simulation champion keeps management informed of developments in the technology and helps identify opportunities for applying simulation within the company. This individual also promotes the technology within the corporation and helps educate potential users in the use and benefits of simulation.

The simulation expertise in an organization need not necessarily be focused in a single individual. It is often wise to have more than one individual involved if for no other reason than to provide a backup in case the principal specialist becomes unavailable. In bringing simulation technology into an organization, a simulation champion has the responsibility of identifying the simulation needs, evaluating available software, and making sure that the appropriate software and hardware are available. On an ongoing basis, the champion acts as a specialist to provide support to individuals in the organization who may require modeling assistance.

Although major corporations have developed one or more simulation modeling groups, most businesses do not have the luxury of dedicating a full time staff to simulation projects. For organizations with little or no experienced simulation users, it may even be best to utilize simulation consultants, at least initially. There are many qualified experts in the field who can provide not only modeling services, but also training services. A consultant with years of simulation expertise will likely charge high fees. However, the benefits of meeting project deadlines and educating your organization while getting your project done could more than pay for the consultant's fees.

A tendency among some managers is to appoint a simulation champion without providing the individual with the needed resources in time and budget for investigating and promoting the technology. "Growing" the technology is given a low priority and must be worked into an already over-demanding schedule. The process of implementing simulation technology could drag on indefinitely with this approach. Another poor practice is assigning personnel who are really not the best suited for the job, simply because the best people are too busy. Some businesses attempt to utilize summer interns or co-op students to implement the technology. Although this approach seems like a cost effective one, it rarely provides the commitment and consistency necessary to be successful. When a co-op student who took a simulation course in college is given a simulation software product and is expected to try it out on a simulation study of an entire factory in two summer months, it is inviting failure. This is especially true if the student is not familiar with the day-to-day operations of the company, or has not taken a formal training course in using that simulation package. Regrettably, it is often either the simulation software or the technology itself that is blamed for the failure.

Step 2: Determine Simulation Needs

Once one or more individuals have been selected to champion the cause of simulation in an organization, they should begin by making an assessment of the organization's needs for using simulation. Consideration should be given to the following questions for both the near-term as well as the long-term:

- How often will simulation be used?
- What applications will it be used for?
- What are the characteristics of the systems to be modeled?
- Who will be the users?
- Who will be the customers?

Each of these questions has important implications for the way in which simulation should be adopted.

Frequency of Use. In some organizations, simulation is used no more than once or twice a year such as when a major change in operations is being considered. In this situation, it may be preferable to hire a consultant to do the modeling. If a purchase of simulation software is made, it should be a product that is easy to use to avoid repeating a steep learning curve with each simulation project. In organizations that plan to utilize simulation more frequently, it may be desirable to involve more than one individual in the use of simulation. Likewise, it may be necessary to acquire several software licenses and possibly even a site license which can usually be obtained at a considerable discount over individual licensing.

Types of Applications. The types of applications will have a significant impact on the software selection as well as who needs to be involved in the modeling projects. If the applications are extremely diverse, it may be desirable to have more than one simulation product to ensure that the right tool is used for the right job. If ongoing "what if" analysis is to be conducted using a single model, such as a scheduling model, it may even be beneficial to have a consultant build a generic model that can be reused over time.

System Characteristics. The characteristics of the systems that will be modeled should have a significant influence on the software that is selected. It is important that the software selected be capable of handling the size, complexity, and issues associated with the systems to be modeled. The next section on selecting simulation software will discuss how to match the right simulation software to a particular set of system characteristics.

Users. The time, skills, and commitment of the simulation users will have a decided impact on the extent to which simulation gets implemented in a company. Simulation experts will have little problem building complex models and using sophisticated simulation tools. It has already been mentioned that simulation is an experimental tool. As such, its effectiveness lies largely in the modeler's ability to identify the cause-and-effect relationships in the system and conduct appropriate experiments.

The recent emergence of easy-to-use simulation software intended for use by engineers and managers rather than the simulation specialist has put simulation within the grasp of the ordinary user who has little, if any, simulation experience. The most effective decision tools are those that can be used by the decision maker. Often, improvements suggest themselves in the very activity of building the model that the decision maker would never discover if someone else is doing the modeling. While an engineer or a manager may not be as capable or as proficient as a simulation "expert," the dividends in increased understanding of the system operation and acquired skill in conceptualizing system designs will pay off in the long run.

Identifying the potential users helps determine the level of training required to enable simulation to be effectively put into use. Easier simulation tools have largely eliminated the need for programming skills, but they will always rely on the modeler's ability to conceptualize the system operation, design an experimental procedure, and interpret the final results. Given the fact that simulation involves primarily data gathering, model building, and experimentation, a modeler will also benefit from having organizational and communication skills, data analysis skills, modeling skills and familiarity with one or more modeling tools, at least a basic background in statistics and design of experiments, and strong analytical skills.

Customers. The customers of simulation are those for whom the simulation is being performed, whether it be management or even the modeler himself. The ultimate measure of success with simulation is whether the customer is satisfied with the results. Therefore, it is important to understand the customers' needs, educate them about the benefits and drawbacks of simulation, and set their expectations so that they are realistic.

Like many other technologies, simulation serves two types of customers: internal and external. Internally, almost every function in an organization is a potential customer. Manufacturing or industrial engineers may need simulation for evaluating new procedures. Operations managers may need simulation for evaluating alternative work schedules. Marketing can use simulation for demonstrating production capacity to a potential contractor. Purchasing may need simulation to evaluate capital expenditures. Maintenance can use simulation for evaluating alternative maintenance policies. Quality can use simulation for evaluating alternative inspection strategies.

Externally, the customers for simulation technology may be suppliers or customers of an organization. Suppliers can use simulation to understand the impact of an organization's delivery requirements on their businesses. Customers can use simulation to see how capable an organization is in meeting promised delivery dates. Banks or financial investors in your organization can use simulation to understand why they should loan you money to build a new facility.

Both the internal and external customers should be involved in understanding the benefits and drawbacks of simulation. Some innovative companies are now utilizing their simulation experts to educate their customers about simulation. For example, Morton International, a major airbag manufacturer, provides simulation expertise and modeling support to three of its suppliers. Sunhealth Corporation, a major healthcare provider, is another example where corporate simulation experts conduct educational training sessions for their hospital personnel.

Step 3: Assess Current Simulation Technology

With advancements in hardware and software technology, this step is essential to making sure that you are taking advantage of the newest developments. It also helps determine your budget and timeline for coming up to speed. Technology is changing so rapidly that you cannot go by how it was done five or ten years ago. Organizations will benefit by using the most up-to-date technology available. Below are some of the sources of information on current simulation modeling and analysis technologies:

- Seminars sponsored by educational institutions.
- Books and videotapes on simulation modeling and analysis.
- Trade shows and software vendor literature.
- Articles published on simulation in trade journals such as *Industrial Engineering, Manufacturing Systems,* and *Managing Automation.*
- Papers presented in conferences and published in the proceedings of The Winter Simulation Conference, IIE Conference, and Autofact.
- Presentations made in local chapter meetings of professional societies.

During the technology assessment process, you should try to find successful users of simulation and ask questions such as:

- How long did it take you to achieve proficiency with simulation?
- What kinds of problems did you encounter in utilizing the technology (not the software package)?
- How much was your total investment?
- Have you achieved the benefits you anticipated?
- What kind of expert help was used in model development and analysis?
- To what do you attribute your success?

Once you have answered these questions, you should prepare a technology assessment report to summarize your findings.

If simulation is not new to your organization, it is a good idea to review the most recent applications of simulation in the firm. This review should address questions such as:

- What specific benefits were realized?
- Were the results implemented?
- Is the person who performed the simulation still with the firm?

Assessing the current state-of-the-art in simulation technology will give you a good idea of what to look for in simulation software. It may even revise some of your initial conclusions about the types of applications that you intend to use it for.

Step 4: Evaluate and Select Simulation Software

Most companies evaluate software without having well-established criteria. Much time is wasted examining software products with little or no guidelines on what to purchase except that it should generally be easy to use, flexible and inexpensive. As a consequence, many organizations get stuck with a product that falls short in meeting the actual needs. Since software selection is such an important step in getting started, this topic will be discussed in greater detail in the next section.

Step 5: Acquire Hardware and Supporting Software

Although hardware decisions were very important up until three to five years ago, this is no longer the case. Due to tremendous advancements in computer hardware technology and significant reduction in prices, hardware is no longer a critical resource in conducting a simulation project. Most major simulation products run on a PC. It is advisable, however, to run simulations on a high performance PC, since large, complex models can sometimes run very slowly.

In addition to the simulation software, one may need other complementary software such as layout design software, flow analysis software, and input and output data analysis software. This software should be identified when assessing simulation needs.

Step 6: Conduct a Pilot Project

The purpose of a pilot project is to go through an actual exercise in simulation. For this project, your application needs to be small yet realistic (a manufacturing cell or a line with ten to fifteen operations producing two to three products). A common mistake made at this stage is to put too much emphasis on building a model. The emphasis at this point should be to go through all the typical steps in a simulation project.

The pilot project should be completed within two to three weeks and findings should be summarized. You should track the time and resources consumed during the pilot study so that you can provide realistic time and budget estimates for the actual simulation study.

Conducting a pilot project is very much like a warm-up exercise. It should be treated as a learning experience as well as a confidence and momentum builder.

SELECTING SIMULATION SOFTWARE

Up until the 1990s, it was most appropriate to classify simulation software into two classes: simulation languages and simulators. Simulation languages were characterized as being flexible, but difficult to use. Simulators, on the other hand, were characterized as being easy to use, but inflexible. Since that time, significant advancements have taken place in simulation product offerings. Some general purpose simulation languages have added specialized modeling constructs making them easier to use, while some of the more specialized simulators have added general programming extensions making them more flexible. In fact, some simulation products that were classified as simulators in the 1980s are now more "flexible" than some of the traditional simulation languages. The most popular simulation tools today combine powerful industry-specific constructs with flexible programming capabilities, all accessible from an intuitive graphical user interface.

Product Types

Today, it is most meaningful to classify simulation products by their application domain. From an application standpoint, simulation products can be categorized as one of the following types:

- General purpose
- Manufacturing oriented
- Material handling oriented
- Service oriented
- Scheduling oriented

Specific products of each type are advertised regularly in trade magazines such as *Industrial Engineering* magazine, which also publishes regular reviews of current leading simulation products.

General Purpose Products. These software products have generic constructs that can be used to model systems of all types (production, traffic, service systems, etc.). Examples are GPSS, SIMAN, MicroSaint, ModSIM and SLAM. Some of these software products have special extensions or built-in constructs for modeling specific components such as material handling systems.

Manufacturing-Oriented Products. These software products are designed for modeling manufacturing systems. Most of these products were classified as manufacturing simulators in the 1980s. Examples are ProModel, WITNESS, SIMFACTORY, Taylor II and ARENA. Some of these products have complete programming capability as well as many built-in constructs for modeling manufacturing systems. Because these products provide an integrated simulation and animation environment, they have become very popular.

Material Handling-Oriented Products. These software products are designed for modeling complex, automated material handling systems such as automated storage and retrieval systems, automatic guided vehicle systems and overhead bridge cranes. Most of these products provide 3-D graphics capabilities that are important for modeling such devices as cranes, robots, etc. Examples are AutoMod and QUEST.

Service-Oriented Products. These software products are designed for modeling service systems. These products provide special constructs for modeling the unique characteristics of service systems, especially as it relates to the human aspect. Examples are ServiceModel, MedModel, and SimProcess.

Scheduling-Oriented Products. Some simulation software vendors have developed finite capacity scheduling modules as extensions to their simulation products. These are FACTOR, AutoSched, and Provisa. These scheduling extensions share data between the simulation and the scheduling module.

Product Evaluation

Frequently software is evaluated on only two or three criteria such as price and ease-of-use. As a result, it is easy to end up owning software that does not meet all of a company's simulation needs. Many of the important things to look for when evaluating simulation software have nothing to do with the software itself, but rather deal with the services offered by the vendor of the software (training, technical support). To make a full and complete evaluation of simulation software, the following considerations should be taken into account:

- Ease-of-use
- Modeling constructs

- Modeling flexibility
- Graphics and animation
- External interfacing
- Statistical capabilities
- Hardware requirements
- Documentation
- Cost
- Quality of support
- Training
- Modeling services
- Upgrades and enhancements
- Other services

Ease-of-use. Many simulation software vendors advertise capability, but are so lacking in usability that it becomes extremely difficult to achieve any productivity with the software. Sometimes users throw up their arms in despair feeling that they will never be able to get a model built and running. Other simulation software vendors have, unfortunately, misrepresented their products by claims that their software is easy to use. The question one must ask those vendors is: "Easy to use for who and compared to what?" A simulation product may be easy to use for a computer scientist in the sense that it is easier to code the model compared to writing a model in a language like C or FORTRAN. Ease-of-use should be gauged primarily by the time and effort required to learn and use the software. Specific features to look for that make simulation software easy to use include the following:

- Modeling constructs that are intuitive and descriptive.
- Model building procedure that is simple and straightforward.
- Use of graphical input wherever possible.
- Input prompts that are clear and easy to follow.
- Context-sensitive help.
- Simplified data entry and modification.
- Automatic gathering of key performance measures.
- Automatic management of multiple experiments.
- Source level debugging and trace features.
- Output reports that are easy to read and understand.

If the modeling language is foreign and terminology is unnatural or unfamiliar, learning and retention of the language become difficult. It helps to have a close correspondence between the modeling terminology and the language typically used in describing the problem domain.

Straightforwardness refers to how intuitive the interface is. Defining downtimes, for example, should be readily apparent from the modeling environment of the simulation software. Brooks (1978) makes a point of the fact that software can be simple, yet not necessarily straightforward, noting that both "simplicity and straightforwardness proceed from conceptual integrity." Conceptual integrity refers to the unity of the product interface—methods and conventions are consistent and everything fits together as one would expect. A simulation language may be simple in that it provides a set of elementary constructs from which models can be created. However, the convoluted and difficult way in which these constructs are combined may not be at all straightforward. Where certain items are, by nature, complicated, there should be easily accessed online help and documentation describing the feature in further detail.

Software that is easy to use provides convenient facilities and mechanisms for defining models. Currently, the trend is graphical user interfaces, such as Windows, that provide easy fill-in-the-blank approaches to model definition, reducing the amount of keyboard entries. Being able to define objects and relationships graphically in addition to textually also makes products easier to use.

The simplification of data entry and data modification is a feature that is often not appreciated until one has been doing modeling for some time and finds that certain tasks, while straightforward and easy to perform, require an excessive amount of tedious effort to complete. A considerately designed product is one in which input requirements are kept to a minimum and model information is kept as efficient as possible. One way to minimize input effort is through the use of macros and subroutines that allow a segment of logic or a series of similar actions to be defined only once.

The ease of using a particular simulation product can be best determined by conducting the following simple experiment:

1. Define a small but representative problem of your own (if you let the vendor define it, it will be tailor-made to fit his product).
2. Invite someone in your company who knows nothing about simulation to participate.
3. Ask your simulation vendor to build a model of that system while your "guinea pig" is watching how the model is built.
4. Now let the "guinea pig" drive the software and try to build the same model while you watch.
5. Evaluate the degree of difficulty in building the model.

This, by the way, has been done by one major corporation to score simulation products and has proven to be a very effective method. Additional questions to ask pertaining to aids that provide ease of use include:

- Does the software have online help? Context-sensitive? Hypertext?
- Does the software have a built-in trainer or tutorial?
- Does the software come with reference models for how to use its constructs?
- Does the software detect syntax errors?
- Does the software provide descriptive error messages and point you to the appropriate section of the model for making corrections?

Modeling constructs. The modeling constructs of a language are the basic building blocks for defining models. If the language has a rich set of constructs for defining commonly encountered situations, the time and effort involved in model building can be greatly reduced. Modeling constructs can be very primitive or quite complex (not to be confused with complicated). Complex constructs allow complex situations to be modeled that would otherwise take days and sometimes weeks to code and debug. Simulation software should provide well-defined constructs for defining at least the following:

- Processing station characteristics
- Resource management
- Entity characteristics
- Entity arrivals
- Processing logic
- Downtime characteristics
- Shift scheduling

Depending on modeling needs, it may also be helpful to have one or more material handling constructs (conveyors, cranes, etc.).

Remember, when evaluating modeling constructs, you are not so much interested in knowing *whether* a particular situation can be modeled (this is modeling flexibility), but rather exactly *what* mechanisms are in place for easily defining the situation.

Modeling flexibility. Simulation products, especially the ones that are domain-specific, provide built-in constructs for quickly modeling complex systems. However, sometimes, the complexity or special nature of the systems being modeled may require programming capability. Modeling flexibility should be evaluated based on the following three aspects of the simulation product:

1. *User-defined elements.* The software should provide the capability to define the needed variable and logic elements for representing special situations.

2. *Access to state and statistical variables.* The software should provide access to the current state or statistics being gathered for purposes of decision making.
3. *Programming language commands.* The software should provide the same basic capabilities of any high level programming language such as C or Pascal. This includes if-then logic, nested logic, looping (While-Do statements), and mathematical and Boolean operators that can be combined in any way desired.

Another very important question to ask regarding flexibility is "Where can these flexible language elements be used?" The software should allow use of these elements wherever a numeric value is used or wherever decision logic is needed. A customized logic routine should be capable of being triggered by any event or time lapse.

Graphics and animation. The graphical representation of a model can be classified into depiction of static background objects, static model objects, dynamic model objects and dynamically-updated status and statistical information. Static background objects include building walls, aisles, columns, etc. Static model objects include machines, queues, transportation paths, etc. Since static background and static model objects do not possess motion characteristics, they are easily depicted. On the other hand, animation and status updating requirements for dynamically changing objects present a more difficult challenge. Dynamic objects include entities (parts, customers, etc.) and resources (people, lift trucks, etc.). Dynamic data displays include counters, gauges, or plots.

Animation can range from simple animated flow diagrams to detailed, pseudo or real 3-D representations. Spending time developing detailed looking graphics should always be weighed against the need to spend time performing the modeling analysis. An embellished 3-D graphical representation can be justified for certain major presentations, as long as the overall objectives of a simulation study are kept in perspective. One should question the benefits of being able to rotate a model in three-dimensional space and view it from any angle, especially if it provides no additional useful information.

Many simulation products permit the importing of CAD developed layouts as the model picture. The fact that a CAD drawing of an entire facility layout is in a computer, however, should not lead to temptations to build a model of an entire factory at once, at least not at the beginning. A simulation model, to be useful, does not need to provide a one-to-one correspondence between the real facility and the model of the facility. Such important factors as the time to build a model, the time to run it, the time to conduct experiments with it, and the time to understand and explain the results must be taken into account before developing a full-scale model of a facility.

Graphics for many simulation products are specified using specially developed modules or animation packages and are defined separate from the simulation. In other products, the animation is defined along with the simulation. In fact, the graphics actually provide a visual approach to defining the model. This approach is easy and intuitive and, when the model is finished, the graphical layout is all defined. Questions to ask regarding graphics include:

- To what extent have graphics been incorporated into the model building process? Can path networks and process flows be defined graphically?
- How realistic do the graphics appear?
- How easy is it to develop the graphical layout?
- Does the software provide capability to import CAD (AutoCad) drawings? If so, how?
- Can icons and clip art be easily imported from other drawing packages?
- Does the software provide a professional graphics editor that allows graphics to be created, edited, and saved in graphic libraries?
- Can graphics be easily copied or moved between graphic libraries?

External interfacing. Most manufacturing companies have databases with routing or scheduling data and many simulation products allow importing data from these databases. However, this process is usually cumbersome and requires programming skills. You should require that the simulation vendor actually show you how importing and exporting are performed rather than simply having them state that it can be achieved.

Sometimes it is desirable to interface with a program or subroutine outside of the simulation model, such as a spreadsheet application or a subroutine written in C or Pascal that would be helpful to have linked to the model logic.

Specific questions to ask regarding the external interfacing capabilities of the software include the following:

- Does the software allow importing data from databases or other applications? If so, how is it done?
- Does the software allow exporting data to a database or other applications? If so, how is it done?
- Does the software allow linking with external applications or subroutines? If so, how is it done?

Statistical capabilities. Although most general purpose simulation languages have excellent statistical capabilities, many new special purpose products lack important statistical functions. The following questions may be helpful in evaluating the statistical capabilities of the software:

- Does the product provide an input data analysis capability for determining distributions?
- How many built-in distributions are supported? Are empirical or user-defined distributions supported?
- Does the program allow for multiple replications? Can selected seed values be automatically reset between replications?
- Does the software provide confidence intervals? Does the software provide other information to help in statistical analysis of output reports?

Hardware requirements. Many simulation products are developed so that they can run on multiple hardware platforms. The advantage of this approach is that models can be portable across platforms. Sometimes there is a loss of performance when software is not written to take advantage of a platform's specific features.With the rapid developments in hardware technology, it is important to make sure that the simulation product takes advantage of future technologies. Questions to ask regarding hardware configuration include:

- Does the simulation software require a special graphics card?
- Are Local Area Networks (LAN) supported?
- Can the software print or plot model layout, model data file, output statistics or graphs?
- Does the software require special compilers? If so how much are they?
- Can the software be run on engineering workstations?
- How much memory (RAM) is needed to run the simulation software?
- What microprocessor is needed to run the software (80486, Pentium)?
- Does the software run on notebook computers?

Documentation. When evaluating software, documentation is perhaps the most overlooked criterion. Yet documentation of the software capabilities, troubleshooting, as well as "how to model" different situations are extremely important. Both online documentation and printed manuals should be reviewed carefully before selecting software. The care and detail taken to provide good documentation are often a reflection of the vendor's interest in providing quality software and services.

Cost. The cost associated with selecting a particular simulation product involves more than the purchase price of the software license. While a software license may cost $10,000, the labor cost for doing a single simulation project can easily end up being two to three times that amount, if the product is limited and difficult to use. When evaluating costs, therefore, the entire cost of doing simulation projects should be considered and not just the cost of the product itself. More expensive products are typically designed to minimize both the learning time as well as the time to model and analyze systems. One should not be mislead, however, into thinking that a higher price tag necessarily means less difficulty. Another important consideration in software costs is the way in which pricing is based. Some vendors price their products in modules. Therefore, you should understand what those modules are and the costs associated with each module before selecting a software product. Other products come as a complete package with everything included. Questions to ask regarding product pricing should include:

- What does the software purchase price include?
- How much is the annual maintenance and upgrade cost?
- How much is training?
- What discounts are offered for multiple purchases, site licenses, etc.?

Quality of support. Regardless of the expertise of the modeler or ease-of-use of the software, most users of simulation are going to require some level of support from the product vendor. Timely, helpful support is critical, especially when under the pressure of meeting a project deadline. The following questions should be helpful in assessing the quality of the support:

- How is technical support provided (telephone, BBS, fax, Internet)?
- How many people are dedicated to technical support?
- Where is the development of the software done?
- Are the developers of the product willing to talk to the end-users?
- How responsive is the vendor to deadlines of the user?
- How close is your nearest authorized representative?
- How competent is your nearest representative?

Training. Effective training can help new users of the software or novices to simulation get off to a good start. Some vendors offer a standard training course that assumes everyone is on the same level. Other vendors have a variety of course offerings that suit individual needs and focus on applications that are relevant to the particular industry of the attendees. Questions to ask regarding training include:

- How frequent are the training courses?
- What levels of training are offered?
- Is industry-specific training available?
- Where are the training courses held?
- Is on-site training available?
- What is the expertise of the trainers?
- How do you get advanced training?

Modeling services. Some vendors have in-house modeling departments that provide modeling services to their customers. Others have consulting services. Modeling services should not be confused with consulting services. That is, a vendor may be very competent in building a model based on a specification, but may not have the expertise to make system design decisions. Modeling services performed by professional modelers often save time and money over doing the work in-house. If, however, the modeling services are not professionally performed, the results can be disappointing. Often the representative responsible for selling modeling services is very competent, but once the project is purchased, someone with lesser credentials ends up being assigned to perform the project. Questions to ask regarding modeling services include:

- Who will be working on the project?
- What are his/her credentials?
- Can the model be used after the consultant is finished with the project?
- What deliverables (model, report, video, etc.) will you get at the end?

Upgrades and enhancements. Software is perhaps the most rapidly changing commodity on the market today. Many of the simulation tools in use today will undergo significant improvements in just the next year. Some simulation tools may even be obsolete within the next year or two simply because they are not being adequately updated. If a simulation vendor does not have an aggressive R&D program, its product will quickly become outdated. Questions to ask are:

- How often does the vendor provide new releases?
- How does the vendor provide fixes to software bugs?
- Are the new releases compatible with the old ones?
- What kind of testing procedures does the vendor use?

Other services. In addition to expected services, there are other vendor services that simulation users have found very useful. Ask the vendor what other services they provide. Specifically, you might want to ask:

- Does the vendor have a bulletin board or an Internet account?
- Does the company provide a newsletter?
- Does the company have a users conference?
- Where and when are they conducted?

A WEIGHTED CRITERIA APPROACH TO SOFTWARE SELECTION

Having defined the criteria for selecting a simulation product based on simulation needs and expectations, the best method for evaluating alternative products against the criteria is using a weighted score selection method. The steps involved in the procedure are essentially as follows:

1. List the criteria to be used for making the product selection.
2. Weight the criteria in terms of their relative importance.
3. Define a scale for scoring each product against each criterion (1 = poor, 2 = fair, 3 = good, 4 = excellent).
4. Obtain product information relative to the criteria that have been identified. If necessary, iterate back to step one to refine the criteria, scoring method, and weight factors.
5. Score features of the software package using the scale and weight factor.
6. Conduct a sensitivity analysis on the results of the evaluation.
7. Select the software with the highest weighted score.

An alternative scoring method when only two or three products are being compared is a *relative ranking* approach in which the best product for each category is identified. The chosen product for each category is then multiplied by the weight factor so that for some categories, the winning product may get double- or triple-counted. The number of entries of each particular product is then added up, with the one having the most entries being the winner.

In the previous section, we provided you with the typical criteria for evaluating simulation software. Table 4.1 shows an example of what an evaluation matrix might look like using the weighted criteria method. The matrix in Table 4.1 evaluates two different simulation products (A and B) based on the fourteen criteria defined previously. For a more extensive evaluation, each of the fourteen criteria could have been further subdivided. Other relevant criteria could also have been included.

Table 4.1 Example Of An Evaluation Matrix.

		PRODUCT A		PRODUCT B	
Evaluation Criteria	**Wt.**	**Raw Score**	**Wt. Score**	**Raw Score**	**Wt. Score**
Ease-of-use	4	4	16	5	20
Modeling constructs	4	3	12	4	16
Modeling flexibility	4	3	12	4	16
Graphics and animation	3	2	6	4	12
External interfacing	2	3	6	1	2
Statistical capabilities	4	4	16	2	8
Hardware requirements	2	2	4	2	4
Documentation	4	3	12	2	8
Cost	4	2	8	1	4
Support	5	2	10	3	15
Training	3	3	9	3	9
Modeling services	2	1	2	4	8
Upgrades and enhancements	4	2	8	2	8
Other services	1	3	3	2	2
Total		37	**122**	39	**132**

For each criterion, a weight is assigned. An easy way to assign weights is to first identify the least important criterion and assign it a weight of 1 (note that if it is determined that a particular criterion is not important at all, it should be removed from the criteria list). In the example, *other services* is considered the least important criterion and is given a weight of 1. Second, the most important criterion is identified and assigned a value based on how many times more important it is than the least important criterion.

In the example, support is considered to be most important and is given a weight of 5 since it is considered to be five times more important than other services. Using these two extremes as a measuring stick, it is relatively easy to assign all of the remaining weights. Weights need not be whole numbers (the lowest weight could be assigned 10, and the highest weight 50, if we wanted to deal strictly with whole numbers).

In scoring products A and B against each criterion, a raw score is first assigned. The raw score is based on a scale from 1 to 4 (1 = poor, 2 = fair, 3 = good, 4 = excellent). A weighted score is then computed by multiplying each raw score by the weight for that criterion. A total score is then calculated for each product by adding all of the weighted scores for each product. Note that the total weighted score for product A is 122 and for product B is 132. This indicates that product B is the preferable selection.

The scoring of products should be done by evaluating each product using questions similar to those provided in the previous section on Product Evaluation. In order to eliminate bias in an evaluation, it may be beneficial to involve two people with each person scoring each product. Obviously each software vendor is developing new products and providing additional features, so the state-of-the-art may change in the future.

Budget Requirements

Investing in a new technology like simulation requires a financial plan and a budget. In order to receive the anticipated benefits from simulation, one must understand the initial costs as well as ongoing costs and make sure that sufficient money is budgeted. Costs include software costs, training costs, hardware costs and user wages. A typical breakdown of these costs by percentage is shown in Figure 4.1.

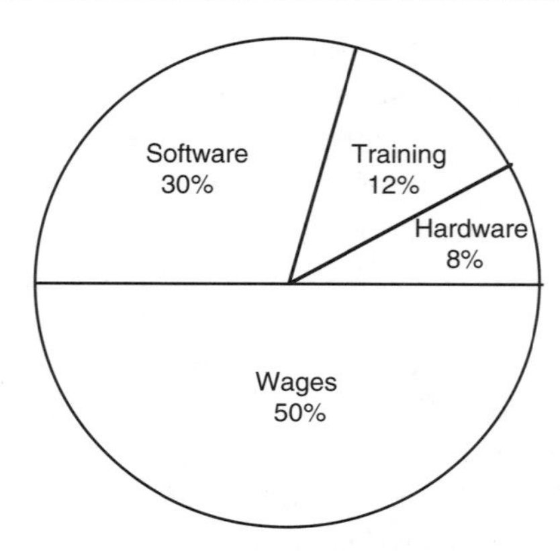

Figure 4.1 Typical Breakdown of Costs For Getting Started in Simulation.

The exact amount to budget depends on the degree of expertise, hardware availability, price and ease of learning of the software. A recommended budget for a single user to get into and continue in the use of simulation is shown in Table 4.2.

Table 4.2 Recommended Budget for a Single User to Use Simulation.

Item	Initial Cost	Annual Cost
Software and support	$5 - $15K	$1 - $3K
Hardware	$3 - $5K	$1 - $2K
Training	$3 - $5K	$3 - $5K
User wages	$10 - $20K	$5 - $10K
Total	**$21 - $45K**	**$10 -$20K**

To understand the nature of these costs, each is examined below.

Software costs. Of course, software costs vary depending on the product. However, the software costs should be divided into initial costs and ongoing costs. Initial purchase costs can be estimated between $5,000 and $15,000 per license and annual maintenance and support costs can be estimated between $1,000 to $3,000.

Training costs. Training costs should also be viewed as initial and ongoing costs. Initial training costs cover general training in simulation and analysis and specialized training in the use of a particular product. This is estimated to be between $3,000 to $5,000. Ongoing training costs are for advanced product training and participation in an annual simulation conference. These costs should be estimated between $3,000 to $5,000 annually.

Hardware costs. Hardware costs should include not only the cost of a computer but also the peripherals (modem, fax). These costs can be estimated at $3,000 to $5,000, with a $1,000 to $5,000 annual allowance for upgrades.

User wages. Most people who have become involved with simulation will agree that the most overlooked cost in a simulation project is time, especially if a company has no previous experience with simulation. We have seen as much as eight to twelve months of engineering time go to waste because of steep learning curves with simulation software or delays in completing the study. Considering an engineer's salary multiplied by an overhead factor, the engineering time for conducting a simulation study can easily reach $80,000 to $100,000. Unfortunately, this is a hidden cost and it is rarely taken into account when budgeting. It is one reason for seeking an outside consultant's help in conducting the initial project.

Time Requirements

It is difficult to estimate exact timelines for various activities in getting started with simulation. Many factors, such as the organization's experience with simulation, size of the organization, and the simulation software used, affect the duration of these activities. However, we provide some loose ranges for the most common processes involved in getting started. Table 4.3 shows tasks and estimated times for preparing to do the first simulation project. In Figure 4.2, these task times are charted on a timeline.

Table 4.3 Time Estimates for Preparing to Do First Simulation Project.

Task	Estimated Time
Assign a person to be responsible for simulation	2 weeks
Become informed on simulation technology	6 weeks
Assess needs and define requirements	4 weeks
Evaluate and select software	12 weeks
Purchase software	8 weeks
Purchase hardware	4 weeks
Become trained in using the simulation software	2 week
Conduct a pilot project	4 weeks

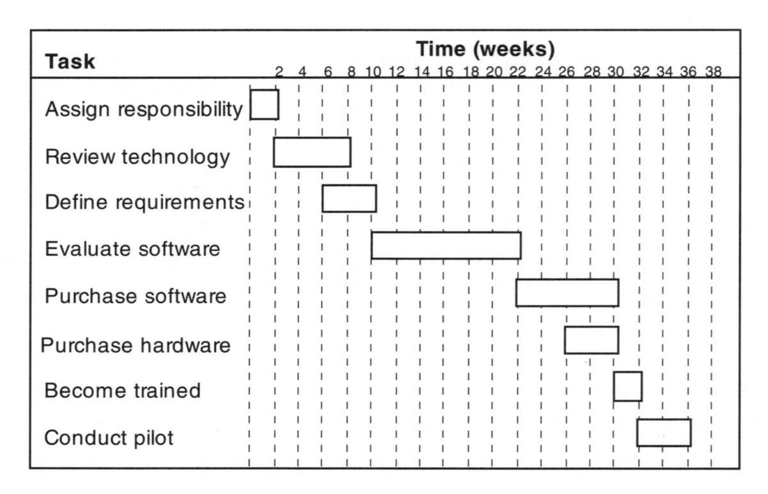

Figure 4.2 Timeline for Preparing to Do First Simulation Project.

Note that some of these activities can overlap, such as defining requirements, which can be performed while reviewing technology. Questions to ask that will affect the budget and schedule include the following:

- Use in-house hardware or software engineers, or outside consultants?
- Use existing hardware or acquire new computer?
- Use corporate simulation group or use project engineers in this facility?
- Single-user or multi-user modeling environment?

SUMMARY

Getting into simulation requires careful planning and preliminary work well in advance of jumping into the first project. In this chapter, we recommended a procedure for getting started in simulation. The steps involved in the successful implementation of simulation technology include the following:

- Appointing a simulation champion.
- Determining modeling needs.
- Assessing current simulation technologies.
- Evaluating and selecting software.
- Acquiring hardware and any additional software.
- Receiving training in simulation modeling and analysis.
- Obtaining simulation software training.
- Conducting a pilot project.

By conscientiously following these steps, many of the pitfalls encountered when first getting started in simulation can be avoided.

Chapter 5
Steps for Doing Simulation

"If you don't know where you're going, any road will take you there."

Unknown

INTRODUCTION

Doing simulation requires more than just knowing how to use a simulation package. Like any project, there needs to be a plan and an understanding of the requirements of each of the tasks involved. Many failures in the use of simulation result from hastily jumping into a project without first taking time to develop a plan of attack.

Despite developments in simulation to make models easier to define, simulate and analyze, there is still much work to be done on the part of the modeler. Hoover and Perry (1990) note: "The subtleties and nuances of model validation and output analysis have not yet been reduced to such a level of rote that they can be completely embodied in simulation software." Simulation modeling requires good analytical, statistical, communication, organizational and engineering skills. The modeler must be able to understand the system being investigated and be able to sort through complex cause-and-effect relationships. A basic foundation in statistics is needed to properly design experiments and correctly analyze and interpret input and output data. Ongoing communication with owners, stakeholders, and customers during a simulation study is vital to ensure that a valid model is built and that everyone understands the objectives, assumptions, and results of the study.

GENERAL PROCEDURE

There are no strict rules on how to perform a simulation study, however, the following steps are recommended as a guideline (Shannon 1975, Gordon 1978, and Law and Kelton 1991):

- Establish objectives and constraints.
- Gather, analyze, and validate system data.
- Build an accurate, useful model.
- Conduct simulation experiments.
- Document and present results.

Each step need not be entirely completed before moving on to the next step. The procedure is often iterative in that each activity is refined, and possibly even redefined, with each iteration. Describing the iterative nature of doing simulation, Pritsker and Pegden (1979) make the following observation:

> The stages of simulation are rarely performed in a structured sequence beginning with problem definition and ending with documentation. A simulation project may involve false starts, erroneous assumptions which must later be abandoned, reformulation of the problem objectives, and repeated evaluation and redesign of the model. If properly done, however, this iterative process should result in a simulation model which properly assesses alternatives and enhances the decision making process.

When doing a simulation project, there are several major checkpoints where a formal review and evaluation should be made of progress. As shown in Figure 5.1, major checkpoints should occur after defining objectives, building the model and conducting experiments.

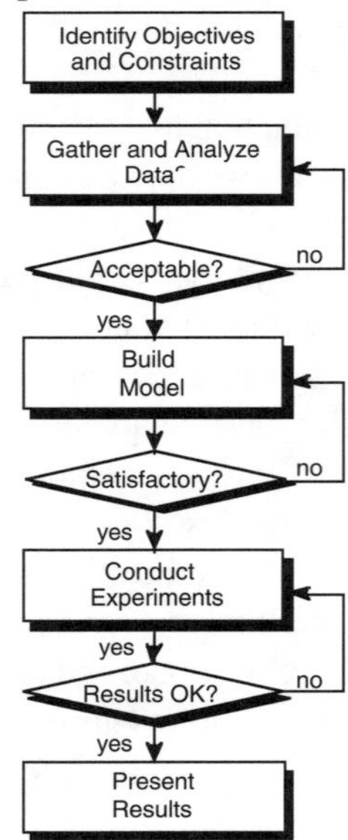

Figure 5.1 Procedure for Conducting a Simulation Study.

The remainder of this chapter provides a more detailed explanation of each of the steps involved in a simulation study and recommendations for how they should be implemented.

IDENTIFYING OBJECTIVES AND CONSTRAINTS

Many simulation projects fail because they get off to a bad start. Misguided expectations, misunderstood directives, and general failure to properly plan the project inevitably result in disappointing outcomes that waste time and resources. If a simulation project is to be successful, objectives must be clearly defined and constraints well understood.

In identifying objectives and constraints, the following questions should be asked (Knepell and Arangno 1993):

- What is the purpose of the simulation?
- Who is the model for in terms of who will ultimately use the outcome of the simulation (the customer)?
- How important are the decisions that will be made as a result of the simulation model?
- What are the customer's expectations?
- What are the budgets?
- What is the deadline?

Appropriate Simulation Objectives

Simulation should only be performed when there are one or more clearly defined objectives and it is determined that simulation is the most suitable tool for achieving these objectives. Defining an objective does not necessarily mean that there needs to be a problem to solve. A perfectly valid objective may be to see if there are any unforeseen problems or merely to gain additional insight into how the system operates. Model building is an analytical discipline that trains one to think logically through the detailed operation of a system. In many instances, the most glaring problems are uncovered and resolved through the model building activity itself before any simulation run has been made. Sometimes a model is built purely as a communication aid to enable others to visualize the complex dynamics of a system.

Common objectives for simulation include the following:

- *Performance analysis.* How well does the system perform under a given set of circumstances in all measures of significance (utilization, throughput, waiting times)?
- *Feasibility study.* Is a system capable of meeting performance requirements (throughput, waiting times, etc.); if not, what changes (added resources, improved methods) are recommended for making it capable?

- *Capacity analysis.* What is the maximum processing or production capacity of a system?
- *Comparison study.* How well does one system perform compared to another?
- *Sensitivity analysis.* Which decision variables are the most influential on overall system performance?
- *Optimization.* What combination of feasible values for all decision variables best meets overall system goals?
- *Decision/response analysis.* What are the relationships between the values of one or more decision variables and the system response to those changes?
- *Constraint analysis.* Where are the constraints or bottlenecks in the system and what are workable solutions for either reducing or eliminating the constraints?
- *Communication.* What is the most effective way to depict the dynamic behavior of the system?

Governing Constraints

Equally as important as defining objectives is identifying the constraints under which the study must be conducted. It does little good if simulation solves a problem if the time to do the simulation extends beyond the deadline for applying the solution, or if the cost to find the solution exceeds the benefit it provides.

Objectives should always be established in light of the constraints under which the project must be performed such as the budget, deadlines, resources available, etc. If no deadlines or other constraints are established, there is a risk of getting too involved and detailed in the simulation study and running the risk of "paralysis from analysis." The scope of the project has a tendency to shrink or expand to fill the time allotted.

A simulation project is subject to all of the same obstacles and hazards of any project. It is generally safe to assume the following:

- The simulation project will probably take longer than anticipated (especially the data gathering phase).
- The simulation project will probably cost more than originally planned (especially when frequent communication may be required).
- The simulation project will require the time of others more heavily than you or they first supposed (coordination and information gathering require lots of cooperation).
- The problems to be solved are more difficult than first thought (some complicated situations may need to be modeled).

- The last 10 percent of the project can be expected to take about 90 percent of the time and will probably end up getting cut short (the time needed to conduct adequate experimentation once the model is built, document the results, and prepare the presentation is rarely budgeted).
- Remember Murphy's Law. If anything can go wrong, it will go wrong. (The corollary to this law is: If something could have gone wrong but didn't, it probably would have been to your advantage if it had!)

Realistic projections must be made of personnel involvement, time requirements for data gathering, experimentation, documentation, and presentation.

DATA COLLECTION AND ANALYSIS

With clearly defined objectives and well understood constraints, the data collection and analysis phase can begin. Data collection should be performed intelligently and systematically to ensure that an appropriate model is built of the system. The process of gathering and validating system data can be overwhelming when faced with the stacks of raw, and often uncorrelated data to sort through. Data is seldom available in the exact form needed for defining a model. Many data gathering efforts end up with lots of data but very little useful information. To help ensure that the data gathering phase provides the proper information for building the model, we recommend that certain guidelines be followed:

- Identify data requirements.
- Use a systematic approach to data collection.
- Use appropriate sources.
- Prepare a list of assumptions.
- Convert data into a useful form.
- Document, review, and approve data.

Determining Data Requirements

Data gathering should never become an end in itself. Remember, the purpose of data gathering is to acquire an understanding of how the system operates. Data gathering should not be haphazard; it should be goal-oriented with a focus on information that will achieve the objectives of the study.

While there may be certain questions that are applicable to nearly every model, each model has unique questions that are specific to the problems being addressed for that particular model. To help create an appropriate list of questions to be answered for a particular model development, we recommend that five steps be followed:

- *Step One.* Determine the required scope (boundary or extent) of the model in light of the objectives to be achieved. In modeling a manufacturing system, for example, you will need to consider whether you should include connecting subsystems such as feeder lines to a final assembly, or to the receiving and shipping operations. This decision should be based on how much impact a particular activity has on the system under study. Do not waste time gathering and verifying data that lies outside of the model scope.
- *Step Two.* Determine the level of detail or resolution required to achieve the model objectives. The level of detail in a model can make a significant difference in the time and difficulty required to build the model. It also impacts how well the model achieves the desired objectives. Too much detail makes the model difficult to build, manage, and analyze and may result in lengthy run times. Too little detail may make the model too unrealistic by excluding critical variables. Do not waste time gathering information on the workstation level if you are modeling the system on a department level.
- *Step Three.* Determine as quickly as possible which factors have a bearing on the objectives. For instance, if downtimes, or move times don't appreciably affect the outcome of the model, they may be ignored safely. Do not waste time getting information on irrelevant factors.
- *Step Four.* Identify, as far as possible, the decision variables to be included in the model. Decision variables are input variables to the model that make sense to modify in order to improve the performance of the model. Decision variables will be needed during the experimentation phase to test alternative solutions.
- *Step Five.* Remind yourself that a simulation is an abstraction of the system, which means you do not model the mechanics, you model essence. The essence of what is happening should be established, not necessarily the mechanics of how things happen.

As these guidelines are followed, it will become apparent what information will be needed to build the model.

To avoid confusion during data gathering, one should carefully distinguish between data used to drive the system (input variables) and data used to describe the system performance (response variables). Data gathering should be aimed at gathering data that actually drives the system and, therefore, can be directly used in building the model. System performance data, on the other hand, is what simulation is intended to provide as an output from running the model. Consequently, data describing system performance should only be gathered to later help validate the model once it is built and run. The amount of work-in-process (WIP) on a shop floor, the average number of customers in a waiting room, entity waiting times, resource idle times, and resource utilizations are all measures of system performance — not input variables that describe how the system operates.

Identifying cause-and-effect relationships.
One of the most difficult aspects of the system to define is cause-and-effect relationships. It is important to correctly identify the causes or conditions under which activities are performed. In gathering downtime data, for example, it is helpful to distinguish between downtimes due to failure, planned downtimes for breaks, tool change, etc., and downtimes that are actually idle periods due to unavailability of stock. Once the causes have been established and analyzed, activities can be defined properly in the model.

Part of the challenge in determining cause-and-effect relationships is to be able to distinguish between scheduled events and conditional events. *Scheduled events* are those that can be defined by a time, even if the time is a probability distribution. *Conditional events* occur when certain defined conditions within the system are satisfied. Conditional events are therefore unpredictable in that their completion times cannot be known *a priori.* An example of a conditional event might be filling a customer order or an assembly operation that cannot be completed until all of the entities to be filled or assembled are available. Often, conditional events can be modeled as scheduled events. If, for example, assembly components are always available for a particular assembly operation, the assembly completion may be treated as a scheduled event activity without specifying the dependency on the entities.

Many events are partially scheduled and partially conditional. Consider for example, a machining operation that requires refixturing midway through a machining cycle. The machining portion of the activity can be scheduled, while the fixturing portion of the activity is dependent on the availability of the operator. When gathering data on activity times, it is important to distinguish between the time actually required to perform the activity and the time spent waiting for resources to become available, or other conditions that have to be met before the activity can be performed. If, for example, historical data is used to determine the repair time for a resource failure, the time spent doing the actual repair work should be used without including the time spent waiting for a repair person to become available.

In grouping multiple activities into a single activity time for simplification, consideration needs to be given as to whether activities are performed in parallel or in series. If activities are parallel, their times should not be additive. If, for example, a part is loaded onto a machine at the same time a finished part is unloaded from the machine, only the longest of the two times needs to be included. Serial activities are always additive. For example, if a series of activities are performed on an entity at a location, rather than specifying the time for each activity, it may be possible to sum activity times and enter a single time or time distribution.

In any simulation study, consideration needs to be given to abnormal behavior or situations that arise infrequently and are not part of the normal operating characteristics of the system. Examples might include:

- A customer in line suddenly becomes ill and leaves.
- Once or twice a month, a part becomes jammed in a machine.

If such behavior is frequent enough to have an impact on the system, they should be included in the model. If however, they occur only rarely, they may be ignored in the normal simulation and dealt with as a "worst case" scenario.

Systematic Data Gathering

Once data requirements have been defined based on the objectives and strategy for building the model, the data collection can begin. Data should be gathered systematically, beginning with general information about the system and then gathering more specific information in a progressive refinement process. This approach provides a framework for information, making it easier to piece the information together. A side benefit of progressively refining the data is that it allows a model to get up and running sooner, which reduces the amount of time to build and debug the model later. Often, missing data becomes more apparent after starting to build the model.

Beginning at the most general level of data gathering, perhaps the best way to get started is to define the flow of entities through the system. An entity flow diagram (sometimes called a work flow or material flow diagram) is an excellent way to visualize the physical flow of entities from location to location. Once a flow diagram is made, a structured walk-through can be conducted with those familiar with the operation to ensure that the flow is correct and that nothing has been overlooked. The next step is to define how entities move from location to location and what resources are used for performing the operations at each location. The final step is to determine the values to use for capacities, move times, processing times, etc. In manufacturing systems modeling, the modeler must often estimate cycle time or downtime data based on time studies, predetermined time standards, historical data, equipment specifications, vendor claims, or one's own best guess. For service systems, log books, time studies or estimates are usually the best source of information. Even though values may be estimated, they can still provide useful insight into system performance.

To direct data gathering efforts and ensure that meetings with others, on whom you depend for model information, are productive, it may be useful to prepare a specific list of questions that lead to the gathering of the right type of information. A sample list of pertinent questions might include the following:

- What are the types of entities that are processed by the system and what attributes, if any, distinguish the way in which entities of the same type are processed?
- What are the route locations in the system (include all locations or points where processing, queuing, or routing decisions are made) and what are their capacities (how many entities can each location accommodate or hold at one time)?
- If entities queue at a location, in what order does the queuing occur (first in, first out; last in, first out, etc.)
- Besides route locations, what other types of resources (personnel, vehicles, etc.) are used in the system and how many units are there of each type (resources used interchangeably may be considered the same type)?
- Where, when, and in what quantities do entities enter the system (define the schedule, interarrival time, cyclic arrival pattern, or condition that initiates each arrival)?
- What is the routing sequence for each entity type in the system?
- What activity, if any, takes place for each entity at each route location (define in terms of time required, resources used, operation logic or other entities that may be involved)?
- If an output entity can be routed to one of several alternative locations, how is the routing decision made (most available capacity, first available location, probabilistic selection, etc.)?
- How do entities move from one location to the next (define in terms of time and/or resources required)?
- What triggers the movement of entities from one location to another (available capacity at the next location, a request from the downstream location, an external condition)?
- How do resources move from location to location to perform tasks (define either in terms of speed and distance, or time)?
- What do resources do when they finish performing a task and there are no other tasks waiting (stay put, move somewhere else, etc.)?
- If there are situations where multiple entities could be waiting for the same location or resource when it becomes available, what method is used for granting access (longest waiting entity, closest entity, highest priority, preemption, etc.)?
- What is the schedule of availability for resources and locations (define in terms of shift and break schedules)?

- What kinds of planned interruptions do resources and locations undergo (scheduled maintenance, setup, changeover, etc.)?
- What kinds of random failures do resources and locations undergo (define in terms of distributions describing time to failure and time to repair)?

Appendix A illustrates how data are gathered using a questionnaire based on these questions.

Depending on the purpose of the simulation and level of detail needed, additional questions may also need to be asked. Answers to these questions will provide the information necessary to build the model. More specifically, they will help define the decisions affecting the delay and flow of entities in the system, how those decisions are made and the conditions in the system that determine the outcome of those decisions.

Sources of Data

With knowledge of what information is required for building the model, data can now be gathered. Data seldom comes from a single source. It is usually the result of tabulation, extrapolation, interviews, and lots of guesswork. "It has been my experience," notes Carson (1986), "that for large-scale real systems, there is seldom any one individual who understands how the system works in sufficient detail to build an accurate simulation model. The modeler must be willing to be a bit of a detective to ferret out the necessary knowledge." Good sources of system data include the following:

- Process plans
- Time studies
- Flow charts
- Facility layouts
- Personal interviews
- Facility walk-throughs

Assumptions

It does not take long after data gathering has started to realize that certain information is unavailable or the data may be unreliable. Complete, accurate, and up-to-date data for all of the information needed is rarely obtainable. For new systems being modeled, completely accurate data is nearly impossible to obtain.

For phenomena where information is unknown or impossible to ascertain, assumptions must be made. There is nothing wrong with assumptions as long as they can be agreed upon, and it is recognized that they are only assumptions. Any design effort must utilize assumptions where clear information is lacking. Often sensitivity analysis, in which a range of values are tested for potential impact, can give an indication of just how accurate the data really needs to be. A decision can then be made to firm up the assumptions or to leave them alone. If, for example, the variance of a particular activity time has little or no impact on system performance, then a constant activity time may be used. Otherwise, it may be important to have a precise definition of the activity time distribution.

Another approach in dealing with assumptions is to run three different scenarios showing a "best case" using the most optimistic value, a "worst case" using the most pessimistic value, and a "most likely case" using a best estimate value. This will help determine the amount of risk you want to take that an assumption will have a particular value.

Assumptions should be documented as such so that they can be approved by the customer before getting too far into the simulation study.

Converting Data to a Useful Form

Data is seldom in a form that is ready for use in a simulation model. Usually, some analysis and conversion need to be performed for data to be useful as an input variable to the simulation. Random phenomena must be either fit to some theoretical distribution or described using an empirical distribution (see Law and Kelton 1991). To define a distribution using a theoretical distribution requires that the data, if available, be fit to an appropriate distribution that best describes the variable (Commonly used theoretical distributions in simulation are given in Appendix B). Several distribution fitting packages (BestFit, UniFit II, StatFit) are available to assist in fitting sample data to a suitable theoretical distribution. The alternative is to summarize the data in the form of a frequency distribution that can be used directly in the model. This is called an *empirical distribution* or a *user-defined* distribution.

Whether fitting data to a theoretical distribution or using an empirical distribution, it is helpful to construct a frequency distribution for the data. Defining a frequency distribution is done by grouping the data into intervals and then stating the frequency of occurrence for each particular interval. To illustrate how this is done, Table 5.1 tabulates the number or frequency of repairs for a particular machine that required a certain range of time to perform.

Table 5.1 Frequency Distributions for Repair Times.

Repair time (minutes)	Number of observations	Percentage	Cumulative percentage
0 - 1	25	16.5	16.5
1 - 2	33	21.7	38.2
2 - 3	30	19.7	57.9
3 - 4	22	14.5	72.4
4 - 5	14	9.2	81.6
5 - 6	10	6.6	88.2
6 - 7	7	4.6	92.8
7 - 8	5	3.3	96.1
8 - 9	4	2.6	98.7
9 - 10	2	1.3	100.0
Total observations = 152			

Note in the last column of Table 5.1 that the percentage for each interval may be expressed optionally as a cumulative percentage. This helps verify that all 100 percent of the possibilities are included.

When gathering samples from a static population, one can apply descriptive statistics and draw reasonable inferences about the population. When gathering data from a dynamic and possibly time varying system, however, one must be sensitive to trends, patterns and cycles that may occur with time. The sample drawn may not actually be a homogenous sample and, therefore, be unsuitable for applying simple descriptive techniques.

Data Documentation, Review, and Approval

When all of the relevant information has been extrapolated or adduced from the data available, it is advisable to document the source of this information and subject it to the scrutiny of others who are in a position to evaluate the validity of the data and approve the assumptions made. Organizing the information into a written summary will make the data easier to review. It will also be useful if you need to come back to reexamine the data or to provide supporting information later to users of the model. In addition to developing a written summary of data, it is also useful to develop a flow diagram of the system operation. This process helps clarify for the modeler his own concept of the real system and helps identify places where information is deficient. It also provides a good conceptual model for how the system operates that can be used as the basis for developing the simulation model.

Validating system information can be a time consuming and nearly impossible task, especially when so many assumptions are made. In practice, data validation ends up being more of a consensus or agreement that is obtained confirming that the information is good enough for the purpose of the model. The approved data summary becomes the basis for the model specification.

BUILDING AN ACCURATE AND USEFUL MODEL

Once the information is at least complete enough so that an assumption list and preliminary model specification can be created, the model building activity can begin. While starting to build a model too early can be a wasted exercise, waiting until all of the information is completely gathered and validated may unnecessarily postpone the building of the model. Getting the model started before the data is completely gathered may even help identify missing information needed to proceed. As mentioned previously, a good model is not one that is necessarily true, but rather one that is useful. A useful model is one that has enough detail and accuracy to meet the objectives of the simulation. The degree to which the model corresponds in detail and accuracy to the actual system is referred to as the model's level of *fidelity*. Higher model fidelity requires longer development, debugging, and run times.

Figure 5.2 shows how dramatically the model development time alone can increase as the degree of fidelity increases. For this reason, it is best to provide no more than the minimum required fidelity to achieve the objectives of the study.

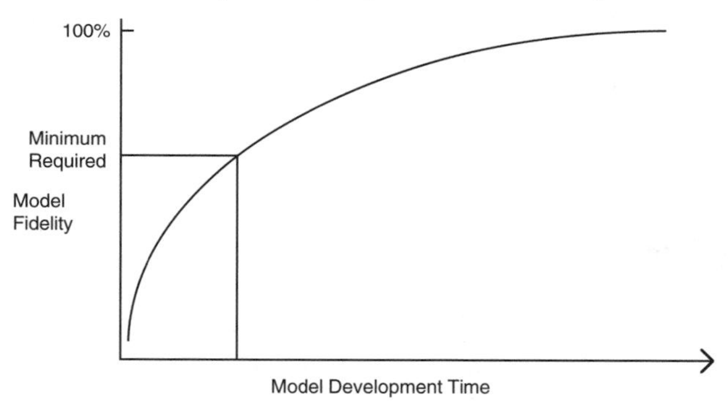

Figure 5.2 Model Development Time to Achieve Minimum Required Fidelity.

For studies in which a process or technological change to an existing system is being considered, it may be useful to model the current system as well as the proposed system. This procedure is recommended for any process improvement or reengineering effort. The basic premise is that you are not ready to make improvements to a system until you understand how the current system operates. Information on the current system is easier to obtain than information on areas of change. Once a model of the current system is built, it is also easier to visualize what changes need to be made for the modified system. Both systems may even be modeled together in the same simulation and made to run side by side. During the final presentation of the results, being able to show both "as is" and "to be" versions of the system effectively demonstrates the impact that the changes can have on system performance.

Model Life Cycle

One issue that should be addressed at the outset of building a model pertains to the life of the model. Questions such as the following should be considered:

- Is it a throw-away model or will it be used again in the future?
- Are there portions of the model that may be useful in the future?
- Who will be using the model in the future?

A tendency has been to view simulation models as throw-away models that outgrow their usefulness once the decisions for which they were developed are made. This trend is changing, however, as increased benefits of simulation modeling are being recognized. For example, simulation may be used to help make strategic decisions at the conceptual design phase. As the design phase begins, modeling of alternative system configurations can be performed. Once a configuration has been selected, the model can be further used at the operational level to find the best control or decision logic for managing the system. The algorithms used can even provide a basis for developing the actual control and information systems specification. In some applications, the simulation model is put to use on a day-to-day basis for production scheduling. A delayed benefit from the model may come later when design changes are made. Since the base model already exists, model changes can be quickly evaluated.

Model Partitioning

It may be useful to identify definable boundaries within a model to permit model partitioning. Model partitioning is the process of breaking down a model into independently operating sections. The purpose of model partitioning is to allow models to be built and debugged, possibly even by separate individuals, independently of each other. Once sections are finished, they can be merged together to create the overall model.

Progressive Refinement

One nice feature of simulation is that models do not have to include all of the necessary detail before they will run. A progressive refinement strategy is usually best where detail is added to the model in stages rather than all at once. This enables a model to get up and running quicker and makes the model easier to debug. In the initial stages of a model, for example, attractive graphics are not very useful and, since they are likely to be changed anyway, should not be added until later when preparing for the final model presentation.

The complexity of model building should never be underestimated and it is always better to begin simply and add complexity rather than create an entire complex model at once. It is easier to add detail to a model than it is to remove it from a model. Building a model in stages enables bugs to be more readily identified and corrected. Emphasizing the importance of applying progressive refinement to model building, Law and Kelton (1991) have advised:

> Although there are few firm rules on how one should go about the modeling process, one point on which most authors agree is that it is always a good idea to start with a simple model which can later be made more sophisticated if necessary. A model should contain only enough detail to capture the essence of the system for the purposes for which the model is intended: it is not necessary to have a one-to-one correspondence between elements of the model and elements of the system. A model with excessive detail may be too expensive to program and to execute.

Model Verification

Once a model is defined using a selected software tool, the model must generally be debugged to ensure that it works correctly. The process of demonstrating that a model works as intended is referred to in simulation literature as model verification. It is much easier to debug a model built in stages and with minimal detail than to debug a model that is large and complex. Eliminating bugs in a program model can take a considerable amount of time especially if a general purpose language is used in which frequent coding errors occur.

Most simulation languages provide a trace capability in the form of an audit trail, screen messages, graphic animation, or some combination of these three. A trace enables the user to look inside the simulation to see if the simulation is performing the way it should. Good simulation products provide interactive debugging capability that further facilitates the debugging process. A thorough "walk through" of the model input is always advisable.

Model Validation

During the process of model building, the modeler must be constantly concerned with model validity. Model validity refers to representational validity and means that the model corresponds to the real system, or at least accurately represents the data gathered and assumptions made regarding the way in which the real system operates. This highlights once again the importance of having valid data at the outset of the model building activity. Proving validity is an elusive undertaking. As Neelamkavil (1987) explains: "True validation is a philosophical impossibility and all we can do is either invalidate or 'fail to invalidate'."

If proving model validity is an impractical, if not an impossible task, then why even make an effort? There are two important points to remember when attempting to prove model validity. First, the more evidence that can be obtained to support the validity of a model, the greater the confidence one has in the outcome of the simulation, hence the greater probability of success of the system. Second, conclusive evidence that a model is valid is unnecessary. The only proof that is necessary is that, for the purposes intended, the model provides a reasonably accurate representation. From this standpoint, model validity can be defined as the process of substantiating that the model, within its domain of applicability, is sufficiently accurate for the intended applications (Schlesinger 1979).

Validation is a rational process in which the modeler draws conclusions about the accuracy of the model based on the evidence available. Gathering evidence to determine model validity is largely accomplished by examining the model structure and evaluating the output statistics. The model structure (the algorithms and relationships) should be checked to see how closely it corresponds to the actual system definition. For models having complex control logic, graphic animation can be used effectively as a validation tool. Finally, the output results should be analyzed to see if the results appear reasonable. If these procedures are performed without encountering a discrepancy between the real system and the model, the model is said to have *face validity.* Face validity means that, from all outward indications, the model appears to be an accurate representation of the system.

CONDUCTING SIMULATION EXPERIMENTS

The fourth step in a simulation study is to conduct simulation experiments with the model. Simulation is basically an application of the scientific method. In simulation, one begins with a theory of why certain design rules or management strategies are better than others. Based on these theories, the designer develops a hypothesis that is tested through simulation. Based on the results of the simulation, the designer draws conclusions about the validity of the hypothesis.

In a simulation experiment, there are certain variables, called *independent* or *input* variables, that are manipulated or varied. The effects of this manipulation on other dependent or response variables are measured and correlated. Because independent variables are the ones that are managed or manipulated in the experimentation, they are sometimes referred to as *supervised variables, decision variables, controlled variables* or *experimental variables*. Since the experimenter is interested in the response of the dependent variables, dependent variables are often referred to as *response* or *performance* variables. As with any experimentation, multiple experiments should be run to test the reproducibility of the results. Otherwise a decision might be made based on a fluke outcome, or at least an outcome that is not representative of what would normally be expected. Simulation languages today often provide convenient facilities for conducting experiments, running multiple experimental replications and evaluating alternative scenarios.

Several types of experiments may be conducted using simulation.

- Finding the expected performance of a particular system design.
- Finding the optimum value for a particular decision variable.
- Finding the optimum combination of values for two or more decision variables.
- Determining the sensitivity of the model to changes in one or more variables.
- Comparing alternative system configurations.

The goal of conducting experiments is not just to find out how well a particular system operates, but hopefully to gain enough insight to know how to improve system performance. Unfortunately, simulation output rarely identifies causes of problems, but only reports the symptomatic behavior of problems. Bottleneck activities, for example, are usually identified by looking for locations or queues that are nearly always full that feed one or more locations that are sometimes empty. This identifies where a likely bottleneck exists. The source of the bottleneck may even be more subtle than identifying the bottleneck itself. Bottlenecks may be caused by excessive operation times, prolonged delays due to a lack of processing resources, or an inordinate amount of downtime. The ability to draw correct inferences from the results is essential to making system improvements.

In conducting any simulation experiment, the modeler must be careful to correctly interpret the output of a simulation run. Since simulation is a vicarious experiment performed on a model rather than the actual system, it is subject to the same types of precautions that should be taken when conducting in any experiment. Since experimental results of a model are random (given the probabilistic nature of the inputs), an accurate measurement of the statistical significance of the output is necessary. In making simulation runs for any particular experiment, the modeler must deal with the following issues:

- Am I interested in the steady-state behavior of the system or is there a specific period of interest?
- What is the best way to ensure that the results reflect only the period of time of interest, without being biased by other conditions such as initial starting conditions?
- What is the best method for obtaining sample observations that may be used to estimate the true expected behavior of the model?
- What is an appropriate length for running the simulation?

- How many replications should be run?
- How many different random streams should be used?
- How should initial seed values be controlled from replication to replication?

The answers to these questions will depend largely on the degree of precision required from the simulation study. People doing simulation in academia are often guilty of working with contrived and often oversimplified assumptions, yet are extremely careful about ensuring the statistical significance of the model results. Simulation practitioners in industry, on the other hand, are usually careful to obtain valid model data, only to ignore the statistical issues associated with simulation output. Maintaining a proper perspective on the statistical aspects of simulation is important to achieving good results. On the one hand, simulation practitioners need to recognize that simulation output is only one sample of many possible outcomes, and that an adequate sample size needs to be taken. On the other hand, modelers who tend to be over preoccupied with the statistical significance need to remember the garbage in, garbage out (GIGO) rule of any software and realize that the ultimate usefulness of the output from the simulation will never be any better than the model input.

The most valuable benefit from simulation is to gain insight, not necessarily to find absolute answers. With this in mind, one should be careful about getting too pedantic about the precision of simulation output. With more than sixty combined years of experience in doing simulation modeling, Conway, Maxwell and Worona (1986) caution that attaching a statistical significance to simulation output can create a delusion that the output results are either more or less significant than they really are. They emphasize the practical, intuitive reading of simulation results. Their guideline is: "If you can't see it with the naked eye, forget it."

Generally, if a rough analysis is all that is needed, a simulation to obtain steady state behavior may be run in which any initial startup bias is omitted from the output through specifying a suitable warm-up period. Then the simulation is run for a lengthy period of time to ensure that there is a clear repeating pattern of behavior in the operation of the model. The resulting average performance values are then used as an adequate estimate of model performance. To obtain suitable results for modeling a specific period of interest, most simulation software provides an automatic replication feature that allows multiple replications to be run automatically while changing the initial seed values for each replication. Running five to ten replications is usually adequate to obtain a good estimate of model performance.

Where more precision is needed or preferred in the simulation output, traditional statistical experiments should be conducted in which confidence intervals can be defined. Chapter 7 on Output Analysis covers traditional statistical methods used in analyzing simulation output.

A final precaution when analyzing the statistical significance of the output of a simulation run or runs is to remember that we are talking only about the statistical significance relative to the model itself and not relative to the actual system. Establishing the statistical significance of the output of a model has nothing to do with the degree to which the output is an accurate predictor of the actual system behavior. To be an accurate predictor of actual system behavior, the model must be valid. Proving model validity is an entirely different issue, which has already been discussed.

DOCUMENTING AND PRESENTING RESULTS

The last step in the simulation procedure is to make recommendations for improvement in the actual system based on the results of the simulated model. These recommendations should be supported and clearly presented so that an informed decision can be made. Documentation of the data used, the model(s) developed, and the experiments performed should all be included as part of a final simulation report.

A simulation has failed if it has produced evidence to support a particular change and then that change is not implemented; especially if it is economically justified. Getting a recommendation to be considered is a selling job. The process of selling a model is largely a process of establishing the credibility of the model. It is not enough for the model to be valid, the client or management must also be convinced of its validity if it is to be used as an aid in decision making. Finally, the results must be presented in terms that are easy to evaluate. Reducing the results to economic factors always produces a compelling case for making a change to the system.

In presenting results, it is important to be politically and personally sensitive as to the potential impact of the decisions influenced by the results. It is important to find out whether recommendations are being sought or whether a simple summary of the results are wanted. It is generally wise to present alternative solutions and their implications for system performance without suggesting one alternative over another, particularly when personnel changes or cuts are involved. In fact, where there may be careers on the line, it is best to caution the decision maker that your simulation study looks only at the operational aspects of the system and that it does not take into account the human and technological issues.

Animation and output charts have become an extremely useful aid in communicating the results of a simulation study. This usually requires that some touch-up work be done to create the right effect in visualizing the model being simulated. In preparing the results, it is often necessary to add a few touch-ups to the model (like a full dress rehearsal) so that the presentation effectively and convincingly presents the results of the simulation study.

After the presentation is finished and there is no further analysis to be conducted (the final presentation seems to always elicit further suggestions for trying this or that with the model), the model recommendations, if approved, are ready to be implemented. If the simulation has been adequately documented, it should provide a good functional specification for the implementation team.

PITFALLS IN SIMULATION

Following the procedure outlined above generally leads to a successful simulation study. However, there are still potential pitfalls that may be encountered by beginners or even by experienced modelers if not careful. Some of the more common pitfalls are listed below:

- Failure to state clear objectives at the outset.
- Failure to involve individuals affected by outcome.
- Naively assuming that simulation, itself, is going to solve the problem.
- Including more detail than is needed.
- Including variables that have little or no impact on system behavior.
- Waiting until completely accurate data is available before beginning the model.
- Basing models on erroneous assumptions or guesses.
- Assuming that the output is valid just because it came out of a computer.
- Basing decisions on a single run without running multiple replications.
- Basing decisions on average statistics when the output is actually cyclical.
- Overrunning budget and time constraints.
- Being too technical and detailed in selling the simulation to management.

SUMMARY

A simulation study, like any other project, has tasks, resources, timelines, and a budget that must be understood. It requires careful planning and coordination to be successful. Steps to performing a successful simulation study include defining objectives, collecting and analyzing data, building and validating the model, conducting experiments, and selling the results.

If the project is incorrectly budgeted and scheduled, the experimentation stage, where the real benefits are to be obtained, can be cut short. Systematically following the steps outlined in this chapter will help modelers avoid the pitfalls that frequently occur when conducting a simulation study.

Chapter 6
Model Building

"If the only tool you have is a hammer, everything looks like a nail."

Abraham Maslow

INTRODUCTION

This chapter examines modeling constructs that are found in modern simulation languages and how to use these constructs to model situations that are common to both manufacturing and service systems. Subsequent chapters will deal with modeling issues that tend to be more industry-specific.

Modeling is considered an art or craft as much as a science. It takes a special knack to be able to mentally transform a system into a simulation model. Developing expertise as a modeler requires an understanding of system dynamics and how to capture these dynamics in a simulation model. It also requires a working knowledge of one or more simulation languages so that systems can be readily translated into model form. Good modeling skills are developed by seeing lots of examples and getting lots of practice. Proficient simulation analysts are able to quickly identify the cause-and-effect relationships in a system and express those relationships using a particular simulation software package.

MODELING PARADIGMS

A simulation model is a computer representation of a system as defined by the software used to develop the model. As we discussed in Chapter 3, all simulation models are ultimately reduced at runtime to a set of events and associated event logic. The primary difference in simulation products is in the way models are defined by the user. Sometimes, and for some individuals, defining a system in terms of events, triggers and associated event logic may not be the most natural or intuitive way of describing a system. Over the years, several modeling paradigms have evolved for defining models. A *modeling paradigm* consists of the set of constructs and associated language that dictate how the modeler needs to view the system being modeled. *Constructs* are the logical building blocks provided by the language from which models are constructed. There are several modeling paradigms or representation schemes used in traditional simulation languages including the following:

- Activity-based paradigm
- Event-based paradigm
- Process-based paradigm
- Object-based paradigm

Rigid paradigms are disappearing as most state-of-the-art simulation products provide modeling constructs that allow the modeler to describe systems in a natural language using familiar terminology. These descriptive constructs are used to define the types of entities processed in the system and the flow of entities through the system. Additionally, most simulation products provide programming capability (if-then logic) for modeling special situations and decision logic.

ABSTRACTING SYSTEM ELEMENTS

A simulation model is an abstraction of the system. To build a model, it is necessary to visualize how the system can be translated into a model using the constructs that are available in the product being used. There are usually several ways in which particular system characteristics can be represented in a simulation model. A resource activity, for example, may be defined using an entity or even a variable if no specific resource construct is used.

Typical manufacturing and service system elements that are often incorporated into simulation models include the following:

- Entities
- Resources
- Movement of entities and resources
- Entity routings
- Entity processes
- Entity arrivals
- Resource availability schedules
- Resource setups
- Resource downtimes and repairs
- Special decision logic

The sections that follow examine how system elements and operational aspects are modeled using modeling constructs that are commonly available in current simulation products. Factors are also presented for consideration when modeling these elements and their behavior.

MODELING ENTITIES

Entities are the items processed through the system (the inputs and outputs of the system). Entities in a system may be of different types and have different characteristics such as speed, size, condition, etc. Entities might get split into other entities or get combined into a single entity. They may arrive from outside of the system or be created from within the system. Usually, entities exit the system after visiting a defined sequence of locations. However, some entities may remain captive to the system such as a container or bin.

Most discrete-event simulation products have an entity construct for modeling system entities. Depending on the product, entities are called *transactions, loads, parts*, or *products*. Nearly every language also permits attributes, sometimes called parameters, to be defined for entities. *Attributes* are like variables that are associated with each individual entity and contain characteristic information about the entity such as size, time in the system, etc. More generic simulation languages even require the modeler to use attributes for defining the entity type.

When deciding what entities to include in a model, it is a good idea to look at every type of entity that has a bearing on the problem being addressed. For example, if products being produced rely on pallets being available for one or more movements, then the pallets should be included as one of the types of entities in the system. On the other hand, if the pallet is always available when needed, and therefore never poses a real constraint, then only the palletizing time may need to be represented, without explicitly modeling the pallet entities.

It is not always necessary nor desirable to define a different type of model entity for each type of system entity. It should be remembered that since a model is generally intended to be a simplified abstraction of the system, it may be preferable to model items with common characteristics as the same entity type. It is a good rule of thumb to only model system entities as separate entity types if there is a significant enough difference in flow or if statistics are required by entity type.

Each system entity need not always be represented by a corresponding model entity. Sometimes a group of system entities can be represented by a single model entity. For example, a single model entity might be used to represent a batch of parts or a group of people eating together in a restaurant.

For high-rate processing where individual item tracking is not critical, it may even be preferable to merely track entity levels using a variable rather than using individual entities. Having lots of entities in the system at one time (especially those with multiple attributes) can consume lots of computer memory and requires extensive processing time which slows down the simulation.

MODELING RESOURCES

Resources are used to directly or indirectly support the processing of entities in the system. In simulation, we are generally interested in how resources are utilized and how entity flow is constrained as a result of waiting for resources to become available. To improve utilization without incurring long waiting times or to reduce entity waiting times without adding more resources, we can experiment with different operating and deployment strategies.

For modeling purposes, we will classify resources on the basis of their functional characteristics:

- Route location resources
- General-use resources
- Consumable resources
- Material handling resources

Some more generic simulation languages make no distinction between different types of resources and model anything "used" by entities the same way. Because of the unique behavioral characteristics of each of these types of resources, however, it is often advantageous to be able to differentiate them in a model. Some languages provide special constructs for modeling each resource type. Each of these classes of resources will be examined individually.

Route Location Resources

Route locations are the places where entities get routed for processing, waiting, or decision making. A route location may be a treatment room, workstation, queue, or storage area. Route locations have a holding capacity and may have certain times that they are available. They may also have special access or admittance rules. The availability of space at a route location may affect whether an entity is allowed to advance forward in the system. Spatial capacity is not always controlled by the actual physical space available, but often by restrictions imposed by management controls. For example, in a health center, no more than one patient may be permitted to enter an examination room, even though more could easily fit. An example of restricting capacity in manufacturing might be the limiting of an input buffer to no more than one pallet load at a time, even though there might be physical space for more.

Most modern simulation products have some sort of construct for modeling the points of routing in a system. Depending on the simulation product, routing locations are known variously as processing locations, workstations, work centers, or buffers. Some languages further distinguish between waiting or buffer areas, such as storages and queues, and the actual processing centers where work is performed. Since route locations are frequently preceded or followed by input or output queues, some simulation languages allow a queue to be attached to a route location to form a single, complex route location. Some simulation languages even permit any combination of elements to be grouped together as complex route locations. In this way, an entire subsystem can be treated as a route location.

Some simulation languages provide constructs for modeling multiple, parallel route locations as a single multi-unit route location (see Figure 6.1). This can simplify model building, especially when the number of units gets very large, such as a twenty-station checkout area in a large shopping center.

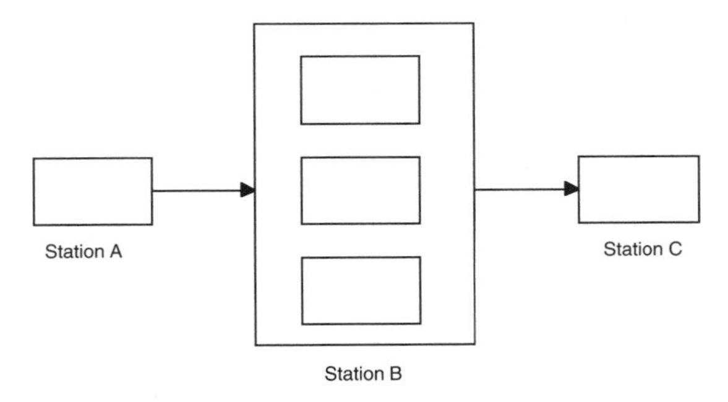

Figure 6.1 Example of a Multi-Unit Route Location.

For modeling simplicity, all points of entity routing will be referred to in a model as route locations. In this way, an entity is always either at some route location, or in transit between route locations.

Determining what to include in a model as a route location is generally quite obvious once the level of detail has been established. Depending on the level of resolution needed for the model, a route location may be used to represent an entire factory or service facility at one extreme, or individual positions on a desk or pallet at the other extreme. In considering what to include as a route location, any point in the flow of an entity where one or more of the following actions take place may be possible candidates:

- Detain an entity for a specified length of time while an activity (fabrication, inspection, cleaning) is performed.
- Detain an entity until one or more resources are obtained.
- Detain an entity until one or more additional entities are joined to it.
- Remove one or more entities from the current entity.
- Consolidate one or more entities into a group.

- Separate an entity into two or more entities.
- Detain an entity until a particular system condition is reached.
- Destroy an entity.
- Create one or more new entities.
- Execute a block of program logic (assignment of variables, etc.).
- Signal the start of some other action in the system.
- Make some decision about further routing (a diverter on a conveyor or a decision point where customers choose among several service counters).

Route locations in a model may have special attributes (cost factors, etc.) associated with them. They also have control rules for selecting among multiple waiting input entities and queuing rules for determining the sequence of entity departure. Since these rules are easily defined in most simulation languages, the modeler need only be cognizant of when such rules are appropriate. A description of these rules is provided below.

Input selection rules. The selection from among multiple waiting inputs is often an important decision to be made, and there are several common rules used to make such decisions. Typical criteria used to select from among several waiting incoming entities include:

- Longest waiting entity
- Shortest waiting entity
- Entity with highest priority
- Entity with highest specified attribute value
- Entity with lowest specified attribute value

Additionally, it may be desirable to be able to select a waiting incoming entity based on some special logic defined by the user.

Output queuing rules. If a location is able to hold more than one entity at a time such as a queue or waiting area, a decision must be made as to the order in which the entities queue to leave the location. The following queuing rules might be defined:

- *No queuing.* Entities are free to leave whenever their individual condition for leaving is satisfied.
- *FIFO (first in, first out).* Entities must leave in the order in which they have entered.
- *LIFO (last in, last out).* Entities must leave in reverse order from which they entered.

- *Maximum attribute value.* Entities should line up to leave, with entities having the highest values for a particular attribute moving to the head of the line.
- *Minimum attribute value.* Entities should line up to leave, with entities having the lowest values for a particular attribute moving to the head of the line.

In addition to commonly-used rules, some simulation languages provide added decision rules that allow entities to renege (leave a queue if waiting time gets excessive) or jockey (move from a long queue to a shorter one).

General-Use Resources

General-use resources include personnel, equipment, and other aids that might be used to process entities, move entities, or service other resources. General-use resources may consist of multiple items of the same type, such as a pool of operators, and may have certain times during which they are available. General-use resources often have specific operating characteristics (movement speed and reliability characteristics) and perform assigned tasks based on either a schedule or priority.

Many simulation languages have a resource construct for modeling general-use resources. As stated in the section above on route locations, some simulation languages have only a generic resource that is passively used. Like entities, resource constructs may permit the use of attributes.

The decisions regarding the modeling of general-use resources usually involve whether to include a particular resource in a model, and the level of detail for modeling a particular resource. The decision of whether a resource should be included in a model depends on whether the model would function measurably differently without using the resource. If the resource is dedicated to a particular workstation, for example, there may be little benefit in including the resource in the model unless the resource experiences downtimes that affect the operation performed at the location.

Sometimes it is necessary to be able to define specific operating policies for a resource. These operating policies may have to do with selecting from among several tasks waiting to be performed. Typical task selection rules are as follows:

- Oldest waiting task
- Closest task
- Highest priority task

Additionally, special decision logic may be needed to define the criterion for performing a particular task.

Consumable Resources

Depending on the purpose of the simulation and degree of influence on system behavior, it may be desirable to model consumable resources. Consumable resources may include any of the following:

- Services such as electricity or compressed air
- Supplies and perishable materials such as staples or tooling
- Money

Modeling consumable resources is usually done indirectly since consumption is a function of either time or observations and nature of the activity. The consumption of packaging materials, for example, might be based on the number of entities processed at a packaging station. Typically, consumable resources are modeled using variables or attributes which monitor and report consumption.

Material Handling Resources

In advanced manufacturing systems, the most complex element to model can be the material handling system. Material handling systems can also pose one of the most difficult modeling challenges due to their complexity and sometimes sophisticated controls. Modeling special material resources can be simplified if special constructs are provided for defining them, otherwise their characteristics must be defined by combining other constructs. The issues to be considered in modeling material handling systems are outlined in Chapter 8.

MODELING MOVEMENT

Entities as well as resources move about in the system. Entities move through the system from route location to route location to have activities performed. Resources also move about in the system to move entities, process entities, or perform maintenance services on equipment. In addition, resources may move to places of their own for maintenance, rest, preparation for a process, or to simply wait until requested.

In deciding how to model entity movement between any two locations, there are several considerations to take into account.

For movement in which no resource is used, the following guidelines are recommended:

- If the move time is negligible compared to activity times, or if the move delay does not hold up downstream processing, it may be ignored.
- If move times are significant *and* entities never encounter traffic *and* the number of different moves in the system is relatively few (less than ten), a simple move time is sufficient for defining the move.

- If significant traffic is encountered *or* the number of different moves in the system is relatively large (greater than ten), a path network should be defined.

For movement in which a resource or material handling system is used (a conveyor system), the following guidelines apply:

- If the move time is negligible compared to activity times *and* the resource is always available, *or* if the move does not hold up downstream processing, ignore the move time.
- If move times are significant and resources are immediately accessible when available *and* resources never encounter heavy traffic, a simple capturing of the resource for the designated move time is sufficient for defining the move.
- If the resource is not immediately accessible when available *or* if heavy traffic may significantly affect travel times, the actual path movement of the resource or operation of the material handling system should be modeled.

Ignoring Movement

Movement can be safely ignored if it is negligible compared to processing times. For example, if each processing step takes twenty to thirty minutes to perform and a move time is only two or three seconds, it probably will not affect the output by more than a fraction of a percent if it is ignored.

In other situations, move times may be significant but performed during off hours, or while the next process already has plenty of entities waiting to be processed. In these situations, any delay caused by the move time has no effect on downstream processing, and can therefore be omitted. An example of this would be a batch of orders that is moved during off-shift hours from one department to another. When processing resumes on the next shift, the move has already occurred.

Sometimes, the level of detail needed for the simulation does not require move times to be modeled, such as when a model is constructed to simulate basic flows. In these instances, the resolution of the simulation may not warrant the inclusion of move times.

Representing Movement Using Time

Time expressions may be used to model simple entity movement where no resource is used. In situations where a resource is used and is immediately accessible, whenever available, simply specify the use of the resource for a designated move time. Consider a situation in which the same operator or clerk performing a process also moves the entities on to the next location.

Since the entity already has possession of the resource, it only needs to reflect the move time for traveling to the next location.

Path Network Movement

For handling complex move patterns or where traffic is a factor, some languages provide the capability to specify path networks, with movement based on speed and distance. If path networks are defined, the shortest path to get between any two points is usually calculated automatically by the software. Paths are used by entities and resources to travel from one location to another. While simply defining a move time from one location to another may work fine for simple models, if the number of locations becomes large, enumerating every move can become laborious since the number of paths between n locations is:

$$\sum_{i=1}^{n}(i-1)$$

or,

$$\frac{n(n-1)}{2}$$

For 25 locations, it adds up to 300 path definitions. This assumes that paths are all bi-directional, otherwise it is much higher. If, on the other hand, the locations are all along a common path, such as the hallway of a building or an aisle of a manufacturing floor, only 25 path segments need to be defined to connect 25 locations. Each additional shortcut or crossover adds only one more segment. In addition to allowing complex path networks to be defined, most high-level simulation software has built-in capability to automatically select the shortest path when moving from one location to another.

Defining a path network provides the added capability to model traffic congestion if that is a consideration.

Material Handling Movement

The *Flexible Manufacturing System Handbook Vol. 5* (1983) cautions against using simple time values to model travel times for material handling systems (MHS):

> If a simulation models the transit times between work stations merely as fixed numbers and the actual MHS is a conveyor, the simulation results may be far from the truth because, depending upon the system's operating policy, parts which cannot enter a work station which is full will either block other parts as it waits or will be "sent around the block." Either operating policy will result in substantial variations in the inter-station transit times which will be unaccounted for in a simple model of the MHS.

Conveyors or transport systems which are complex have a move time that is a function of speed, distance, traffic, selected route. It is often difficult to predict what the move time will be since it is dependent on the state of the system. Such move times are, themselves, determined through simulation of the resource as a subsystem of the overall system. More on material handling movement will be presented in Chapter 8.

MODELING ENTITY ROUTINGS

Entity routings define the sequence and selection criteria for going from one route location to the next in the system. The following information is usually needed to define an entity routing:

- The output entity or entities resulting from the process.
- A list of possible locations to route to.
- The rule or criteria used for selecting the next location.
- The priority for accessing the next location.

Output Quantities

In modeling the input/output relationship at a location, the modeler needs to recognize that there is not always a one-to-one ratio. Sometimes the output quantity can be greater or smaller than the input quantity. Batching, assemblies, and entity splitting all require the modeler to carefully ensure that the right amount of output entities result from a given input entity or set of input entities.

Rules for Selecting the Next Location

If multiple output destinations exist, they each need to be defined along with the rule for selecting one of them. Typical rules that might be used for selecting the next location in a routing decision are as follows:

- *Probabilistic.* Entities get routed to a particular location a certain percentage of the time.
- *Deterministic.* Entities always go to the location(s) specified.
- *First available.* Entities go to the first available location specified in a list.
- *By turn.* Entities go to available locations by turn or in a rotating fashion.
- *Most available capacity.* Entities select from among a list of locations based on which one has the most available capacity.
- *Until full.* Entities go to a single location until it is full and then switch to another location where they continue to go until it is full and so on.
- *Least available capacity.* Entities select from among a list of locations based on which one has the least available capacity.
- *Random.* Entities choose randomly from a list of locations.
- *User condition.* Entities choose from among a list of locations based on a condition or decision logic defined by the user.

In deciding which selection rule best represents the routing decision, the modeler needs to carefully analyze the nature of the routing decision. Sometimes the selection is mutually exclusive, in that a decision is made among several alternative locations. The rules for making this type of selection are fairly straightforward. At times, however, the selection may be all inclusive, in that all of the locations specified receive one or more outputs. An example of the latter is a customer placing an order in a fast food restaurant. After placing the order, the customer moves to a waiting area while an order created by the process gets routed to the order filling area.

Compounding the complexity of making a routing decision are situations where there are multiple outputs, and each output has a different set of multiple destinations from which to choose. Ideally, we would like to be able to define some type of block of alternative destinations for each output.

Access Priorities

Access priorities determine the order in which a given entity waiting for the next location is permitted to "claim" that location when it becomes available. Obviously, an access priority is only meaningful when there is the possibility of two or more entities waiting for a location to become available. For example, both a new part and a reworked part may be routed to the same inspection location. In this case, the modeler may want to give a higher priority to the reworked part.

The modeler should be careful to designate the appropriate access priority if it applies or the flow of entities in the model may not reflect the actual flow of the system. Priorities may sometimes even be preemptive: the entity actually has permission to bump another entity or activity at a location in order to get immediate access to the location.

In a manufacturing system, for example, parts may get routed through a cleaning area or inspection station after more than one operation.

Recirculation

Sometimes entities recirculate back through one or more locations. Some languages accommodate recirculation where others require the use of attributes to track which pass the entity is currently on.

Nonsequential Routing

There are some systems that do not require a specific order for visiting locations, but allow some flexibility so that some activities may be performed in any order as long as they all get performed. In this situation, it is important to keep track of which locations have been visited or which locations remain to be visited. Entity attributes are usually used for tracking this information.

MODELING ENTITY PROCESSES

An *entity process* defines what happens to an entity when it enters a location. For modeling purposes, the exact nature of the process (machining, patient check-in, etc.) is irrelevant. What is essential is to know what happens in terms of the time consumed, the resources used, and any other logic that impacts system performance. For processes requiring more than a time and resource designation, detailed logic may need to be defined using if-then statements or action statements.

An entity process is one of several different types of activities that take place in a system. As with any other activity in the system, the decision to include an entity process in a model should be based on whether the process impacts system performance. Obviously, if a labeling activity is performed on entities while in motion on a conveyor, the activity need not be modeled unless there are situations where the labeler experiences frequent interruptions.

It is often desirable to represent batching, assembly, merging, grouping or other types of consolidation processes in different ways. Most simulation languages provide constructs that allow different ways to combine entities to represent these processes. Depending on the way in which entities are combined, an activity might be defined as:

- Merges or consolidates two or more entities into a new output entity (Figure 6.2).
- Groups two or more entities into a single grouped entity (Figure 6.3).
- Permanently joins or merges two or more entities to another entity (Figure 6.4).
- Temporarily attaches one or more entities onto another entity (Figure 6.5).

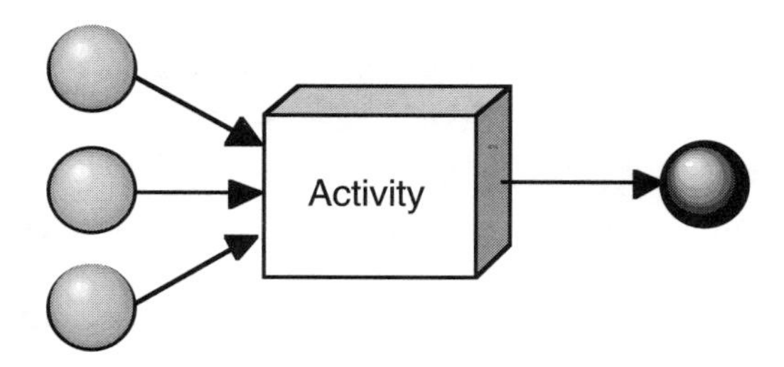

Figure 6.2 Merging Three Entities to Create a New Entity.

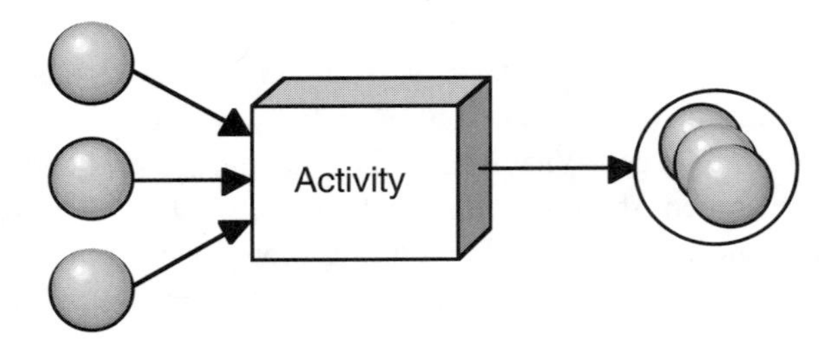

Figure 6.3 Grouping Three Entities Into a Single "Grouped" Entity.

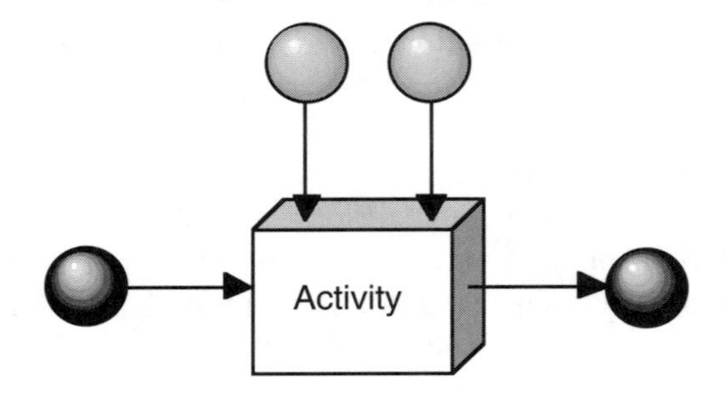

Figure 6.4 Joining Two Entities to Another Entity.

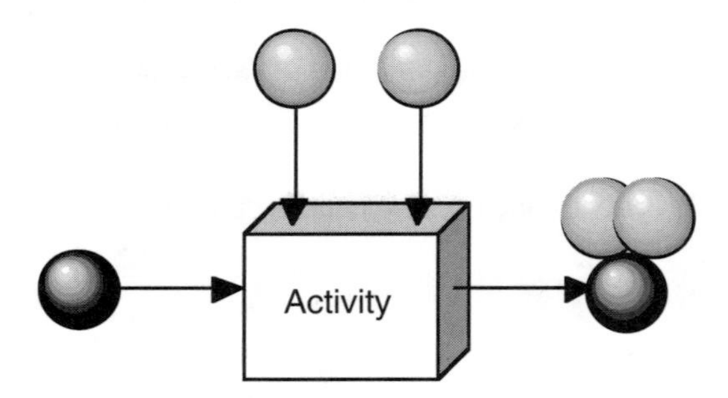

Figure 6.5 Temporarily Attaching Two Entities to Another Entity.

Some processes convert a single input into multiple outputs such as separating sections of a report or unloading boxes from a pallet. When modeling this type of process, it is important to consider the way in which the division occurs. For example, an activity may:

- Split an incoming entity into two or more new entities (see Figure 6.6).
- Ungroup a previously grouped entity (see Figure 6.7).
- Create one or more additional entities (see Figure 6.8).
- Remove entities that have been previously attached. (see Figure 6.9).

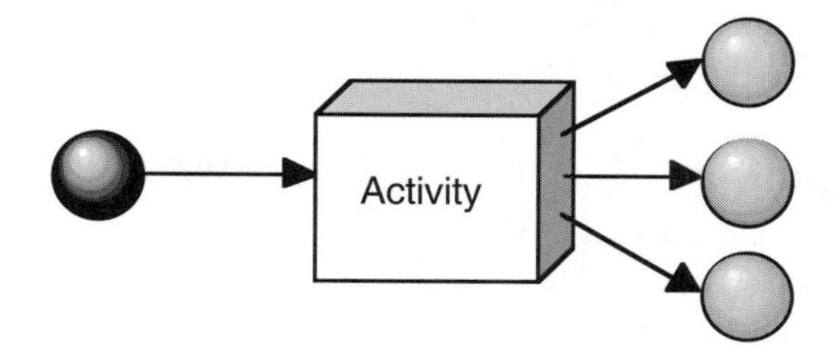

Figure 6.6 Subdividing an Entity Into Three New Entities.

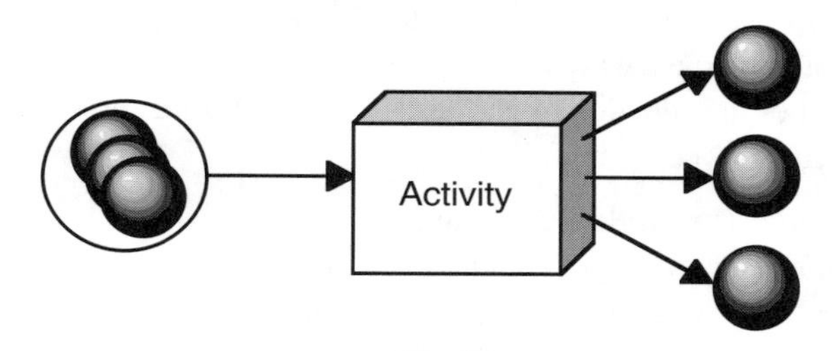

Figure 6.7 Ungrouping an Entity That Has Been Previously Grouped.

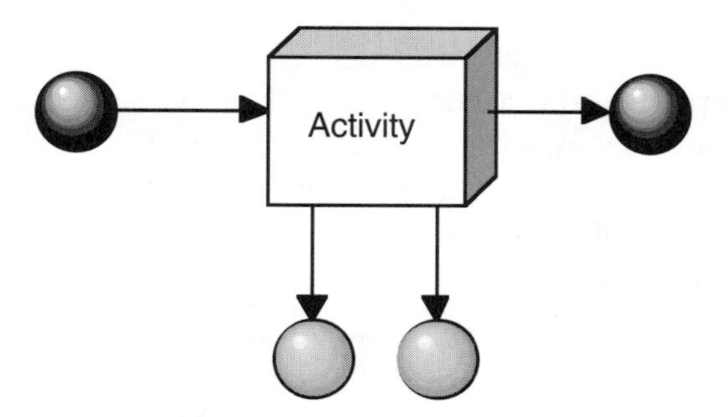

Figure 6.8 Creating Two Additional Entities.

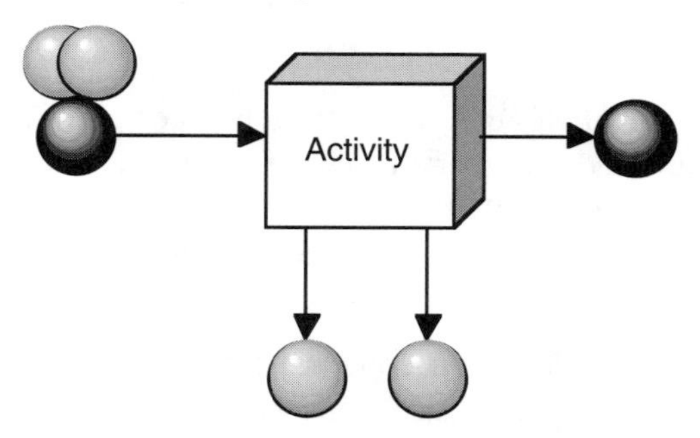

Figure 6.9 Detaching Entities That Have Been Previously Attached.

MODELING ENTITY ARRIVALS

Entities arriving into the system are usually defined based on the following information:

- Entity type
- Route location where it arrives
- Quantity of entities included in each arrival
- Number of arrivals
- Frequency or pattern of arrivals

Since an arrival is an event, it is also useful to permit any user-defined logic to be associated with each arrival.

The arrival or introduction of entities to manufacturing and service systems typically occurs in one of the following ways:

- *Scheduled.* They follow a schedule.
- *Periodic.* They recur at periodic intervals.
- *Cyclic.* They occur according to some cyclic pattern.
- *Internally initiated.* They are triggered by some event.

Scheduled Arrivals

Scheduled arrivals occur when entities arrive independently of one another according to a predefined schedule. Scheduled arrivals may occur in quantities greater than one, and may occur at either an exact time or a probabilistic time. It is often desirable to read in a schedule from an external file, especially when the number of scheduled arrivals is large and the schedule may change from run to run. Examples of scheduled arrivals include the following:

- Customers' appointments to receive a professional service (counseling).
- Patients scheduled for lab work.
- Production release times created through an MRP system.

Periodic Arrivals

Periodic arrivals are arrivals that occur at regular intervals. They may occur in varying quantities and the interval may be defined as a random variable. Periodic arrivals are frequently the result of the output of some other system that feeds into the system being investigated. Examples of periodic arrivals are as follows:

- Parts coming from a machine operating at a fixed cycle.
- Customers arriving to use a copy machine.

Cyclic Arrivals

Cyclic arrivals are similar to periodic arrivals in that they are recurring. They are different, however, in that the pattern of arrivals fluctuates over time. The interarrival time distribution in a cyclic arrival is said to be nonstationary. The fluctuation usually follows a certain pattern that may start out slow at the beginning of a cycle, reach a peak part way through the cycle, and then taper off towards the end of the cycle. Examples of cyclic arrivals include:

- Customers arriving at a restaurant.
- Arriving flights at an international airport.
- Arriving phone calls for customer service.

Internally Initiated Arrivals

In many situations, entities are introduced to the system by some internal "trigger," such as the completion of an operation or the lowering of an inventory level to the reorder point. Examples of internally initiated arrivals include:

- A Kanban system.
- Reorder point inventory management (make-to-stock production).
- Initiation of a customer order.

MODELING RESOURCE AVAILABILITY SCHEDULES

Resources, including route locations, may have scheduled times during which they are available for use. Shift changes, breaks, or scheduled maintenance on equipment make it necessary to cause certain designated resources to become unavailable. Constructs for modeling resource availability usually include the following information:

- The time periods during which the resource is available.
- The rule specifying whether current activities should be completed or interrupted.
- Any logic associated with the start or end of the period of availability.

For many types of systems, it is unnecessary to model off-shift or scheduled maintenance time, especially if the entire system is dormant during this time. If, however, only part of the system is shut down while another part continues to operate, it is usually necessary to model the period of time that only part of the system is functioning.

MODELING RESOURCE SETUPS

After a process occurs, resources and route locations sometimes must undergo some setup activity to prepare for the next entity to be processed. A setup might be a tool change in a manufacturing system or cleaning a table before seating another customer in a restaurant. Resource setup is an activity that may need to be defined as a time or as an action. Furthermore, the amount of setup time may be dependent not only on the current entity to be processed, but also on the preceding entity type. Information required to define a resource setup includes the following:

A setup usually occurs when a new entity type is to be processed, although it may be required for every entity entry. If every entity requires a setup activity, it is easiest to model the setup as part of the processing activity, assuming the same effect can be achieved.

The duration and other requirements of the setup activity are sometimes a function of the entity type to be processed. It may even be partially or wholly determined by the setup that was performed for the previous entity type. Some simulation tools provide the capability to either directly or indirectly specify setup based on the incoming entity type as well as the entity type that was just processed.

MODELING RESOURCE DOWNTIMES AND REPAIRS

Downtimes are the interruptions that occur to a resource usually from a failure or a rest break. Downtimes can be based on elapsed time of the clock, time in use, or the number of times used. Downtime constructs usually provide a way to specify the following:

- Resource name.
- Basis for going down (elapsed time, time in use, number of times used).
- Frequency of the downtime occurrence using the basis for going down as the measure.
- The time and any resources required for making repairs.

Resource availability is the time that a resource is operable and therefore usable. It is uncommon for resources to always be online in a manufacturing facility. Resources are frequently down due to either planned or unplanned interruptions. Unplanned downtime such as machine failures may occur after a specified time in actual useor after a specified lapse of time on the clock. Planned downtimes are based either on clock time as in the case of shift availability or machine cycles such as for periodic tool changes.

Unfortunately, data is rarely available on downtime characteristics of equipment. When it is available, it is recorded as overall downtime and seldom broken down into any frequency distribution, much less reflecting actual repair time.

A typical way of treating downtime, due in part to the lack of downtime characteristics, is to simply reduce the production capacity of the machine by the downtime percentage. In other words, if a machine has an effective capacity of 100 parts per hour and experiences a 10 percent downtime, the effective capacity is reduced to 90 parts per hour. What this essentially does is to spread the downtime across each machine cycle so that both the mean time between failures and mean time to repair are very small and both are constants. Thus, no consideration is given for greater values and variability in both mean time between failures (MTBF) and mean time to repair (MTTR) that typify most production systems. Law (1986) has shown that this deterministic adjustment for downtime can produce results that differ greatly from the results based on actual breakdowns of the machine.

The *availability* of equipment is a function of both reliability and maintainability. Reliability is defined as the probability that a given component or system will perform without failure for a given period of time. *Maintainability,* on the other hand, is defined for a time period as the probability of a component or system being capable of being repaired within that time. Reliability in manufacturing is generally expressed in terms of *mean time between failures (MTBF).* It is calculated as follows:

MTBF = Total operating time / Number of failures

Maintainability is expressed in terms of the *mean time to repair (MTTR).* MTTR is calculated as shown below:

MTTR = Total down time / Number of failures

The sum of MTBF and MTTR is the total scheduled time during which the resource should be available to perform work. The actual percentage of time that the resource is available is termed *availability* and is defined as follows:

Availability = MTBF / (MTBF + MTTR)

From a simulation standpoint, several considerations must be taken into account when modeling random downtimes. First, downtime events or interruptions should not be modeled simply in terms of MTBF and MTTR. Doing so would not produce the desired effect of short downtime events; neither would it reflect the impact of long downtime durations. Time between failures and time to repair should be represented by a probability distribution that is representative of the variability of times that is likely to occur.

Another important consideration in downtime modeling is whether to base MTBF on elapsed time on the clock time or on actual operating or productive time. Some downtimes should be based on clock time such as power failures and unscheduled personal breaks. Often, however, downtimes are based on operating time such as tool breakage and mechanical failures that are only likely to occur while operations are being performed.

Erroneously basing downtime on elapsed time when it should be based on operating time, produces results that are not reflective of the actual system performance. It implies that during periods of high equipment utilization, the same amount of downtime occurs as during low utilization periods. Equipment failures should generally be based on operating time and not on elapsed time since elapsed time includes both operating time and idle time. It should be left to the simulation to determine the effect of any idle time on the overall elapsed time between failures. To illustrate the difference this can make, assume that the following times were logged for a given operation:

Status	Time (hrs)
Productive	20
Down	5
Idle	15
Total time	**40**

If this operation is modeled with downtime expressed as a function of total time, then the percentage of downtime is calculated to be 5 hrs/40 hrs or 12.5 percent. If, however, it is assumed that the downtime is a function of productive time, then the downtime percentage would be 5 hrs/25 hrs or 20 percent.

Modeling Repair Time

Often the repair time can be modeled without regard for resource requirements. If there is a possibility that several activities might be in competition for the same maintenance resource, then it is advisable to model the resource. Otherwise, specifying only a repair time should be sufficient. If historical data is used to determine repair times, it is important to exclude the time spent waiting for maintenance resources when specifying the repair time in the model.

In modeling downtime, it is important to distinguish between downtime due to failure, planned downtimes for breaks, tool change, etc., and downtime that is actually idle time due to unavailability of work. In addition, repair times should be identified by the actual time repairs were made and not on how long it took before a repair person became available.

Handling Interrupted Entities

When a resource goes down, there might be entities that were in the middle of being processed that are now left dangling. For example, a machine might break down while running the seventh part of a batch of twenty parts. The modeler must decide what to do with these entities. There are several alternatives that may be chosen, and the modeler must select which alternative is the most appropriate:

- Leave the entity alone until the resource that went down becomes available.
- Find another available resource to continue the process.
- Scrap the entity.

If the entity resumes processing later using either the same or another resource, a decision must be made as to whether only the remaining process time is used or if additional time must be added.

MODELING SPECIAL DECISION LOGIC

Many decisions that take place in a system have peculiarities that do not allow them to be easily defined using built-in modeling constructs. In these situations, it is necessary to use programming logic that makes use of user-defined data elements. Most simulation products provide variable and attribute data elements for carrying information and making decisions. These data elements are symbolic names that represent a value or perhaps even contain text.

Variables

Variables are generally global in nature, which means that they may be accessed for modification or testing anywhere and at any time during the simulation. Uses of global variables include:

- Keeping track of the number of entities in a particular portion of the system.
- Recording when some event took place.
- Accumulating the number of times a particular routing took place.

Some variables, called *arrays,* are actually lists of variables used for storing multiple data elements of the same type. An array might consist of a list of the types of tools that are currently in use at a location.

Some simulation languages provide local variables in addition to global variables. *Local variables* are like temporary entity attributes in that they are associated with each entity, but only during a defined section of logic. Local variables are useful for providing a loop counter (*while LocalVar1 < 3 do . . .*).

Attributes

Attributes may be thought of as variables associated with a system element such as an entity or resource. Attributes might carry such information as size, weight, speed (if self-moving) or cost. The actual attributes defined depend on the information required to fulfill the purposes of the simulation. A simple model may require no attributes at all.

Use of Programming Logic

Although most modern simulation software products provide flexible built-in constructs, there are inevitably going to be complex situations that require the use of programming logic. Statements such as *if-then-else*, *while-do* are provided in most simulation languages to control complex decision making during the simulation. An example of the use of an if-then-else statement in operation logic might be to use *operator A* for 1 minute if the attribute called *Size* is greater than 20, otherwise *operator B* is used for .5 minutes. Depending on the language constructs provided by the software, the logic for implementing this decision might look something like the following:

```
IF Size > 20
THEN use Operator A for 1 min
ELSE use Operator B for .5 min
```

Interfacing with External Data and Code

Generally, simulation products provide all of the needed data elements and logic capabilities for defining most systems. Occasionally, however, it may be more convenient to access data or subroutines external to the model. Most modern simulation languages provide facilities for accessing external databases, spreadsheet files, etc. They also permit models to be linked to externally programmed subroutines written in C, C++, Pascal, etc.

SUMMARY

Model building requires a knowledge of the language constructs of the particular simulation product being used, and the ability to express the system operation in terms of those constructs. Many currently available simulation products provide intuitive constructs that utilize familiar terminology. The challenge of the modeler is to know what to include in the model and how each particular system element should be represented. The key to success in modeling is seeing lots of examples and practice, practice, practice!

Chapter 7
Output Analysis

"The method that proceeds without analysis is like the groping of a blind man."
Socrates

INTRODUCTION

The purpose of this chapter is to explain how to interpret simulation results. Remember that simulation, by itself, does not optimize a system, nor does it provide solutions. This is the work of the modeler and the user of the model. Simulation merely evaluates solutions by providing estimates of how a given system would behave. To draw useful and correct conclusions from simulation results requires that reliable and relevant data be generated by the simulation, and that the data be correctly analyzed.

What to look for in simulation output depends on the nature of the simulation and the answers being sought. If the model is used only to conceptualize the system dynamics, the animation itself may be sufficient. Where transient behavior is being analyzed, each individual time period during the simulation may be of interest. For steady-state analysis, the startup period is discarded and we are only interested in the long-term behavior of the system.

Statistical analysis of simulation output is generally required when we are trying to perform one of the following types of experiments:

- Finding the expected performance of a particular system design.
- Finding the optimum value for a particular decision variable.
- Finding the optimum combination of values for two or more decision variables.

- Determining the sensitivity of the model to changes in one or more variables.
- Comparing alternative system configurations.

SIMULATION EXPERIMENTS

In studying the topic of simulation output analysis, one can be quickly overwhelmed by the number of different and complex statistical issues associated with simulation output. It should be realized that, as in any type of experiment, there is room for rough analysis using judgmental procedures as well as more statistically-based procedures for more refined and precise analysis. The decision as to how statistically precise one should be in using simulation depends largely on the nature of the problem, the importance of the decision and the validity of the input data. If you are doing a go/no-go type of analysis in which you are trying to find out whether a system is capable of meeting a minimum performance level, then a simple judgmental approach may be adequate. Finding out, for example, if a service facility can get by with only one service representative, may be easily determined by a single run unless it looks like a close call. Even if it is a close call, if the decision is not that important (perhaps there is a backup who can easily fill in during peak periods), then more detailed analysis may not be needed. Similarly, it does little good to get three decimal places of precision in an answer if the input data being used is only accurate to the nearest tens place. Suppose, for example, that incoming customers to a service facility are only roughly estimated at thirty, plus or minus ten, per hour. In this situation, it is probably meaningless to try to calculate the precise utilization of the service representative. The input data simply does not warrant any more than a ballpark estimate.

These examples are in no way intended to minimize the importance of conducting statistically sound and intellectually responsible experiments, but rather to emphasize the fact that a practical engineer or manager can gainfully use simulation without having to be a professional statistician. Anyone who understands the cause-and-effect relationships of a system and can describe patterns of behavior in the system, will be able to draw meaningful conclusions about a system that will help in design and management decisions. Where precision is important and warranted, however, a statistical background or at least consulting with a statistical advisor will help avoid making wrong inferences from a simulation experiment.

TYPES OF OUTPUT REPORTS

There are many different ways in which output data is presented and it helps to know the different types of output reports that are typically available in most simulation software. Simulation output reports can be categorized as follows:

- Single Run Summaries
- Detailed Histories
- Snapshot Reports
- Multiple Replication Summaries
- Multiple Scenario Comparisons

Single Run Summaries

A single run summary shows the performance behavior of a simulation run using mean and variance statistics to indicate system performance. The information included in a single run summary might consist of average activity times or average number of entities in a waiting area.

Detailed Histories

Detailed histories are data logs and time plots showing individual observations or state changes over time. Plots might include such measures as the number of entities in a waiting area or the utilization of a resource over time. This type of output is called time series, time persistent, or time dependent statistics. Such information can be averaged or grouped to provide summary information. For example, a histogram might be constructed showing the time entities spent in the system.

Snapshot Reports

Snapshot reports are generally statistical summaries provided at points of time during the simulation run. Like a balance sheet for a corporation, it shows the status of the system at some point in time.

Multiple Replication Summaries

Multiple replication summaries show the combined results of multiple replications and are therefore able to show an expected mean and standard deviation for the model. Estimates of error based on confidence intervals are also provided.

Multiple Scenario Comparisons

Multiple scenario summaries compare the output of different variations on a system design. If multiple replications were run for each scenario, a confidence interval may also be included.

OBSERVATIONAL VERSUS TIME-WEIGHTED OUTPUT

There are primarily two types of output data reported from a simulation run: *observational data* and *time-weighted data*. Observational data reports the number of occurrences of an incident and has the following characteristics:

- The observations are counts of occurrences.
- The observations are equally weighted.
- The summary statistic is the number of observations over some period of time.

An example of an observational statistic would be the average number of entities processed per hour through the system. This would be calculated by counting all of the entities that exited the system during the simulation, and then dividing that number by the number of hours in the simulation.

Avg. entities per hour = Total entities processed / Simulation hours

Another example of an observational statistic is the average time entities spend in the system. To track the time entities spend in the system, a time stamp is placed on the entity when it enters the system. When it leaves the system, the time stamp is subtracted from the current clock time and the observed time lapse is added to the cumulative amount of time. To compute the average time in the system, divide the cumulative time spent in the system by the number of entities that exited the system.

Avg. time in system = Cumulative time in system / Number of exits

Time-weighted data, on the other hand, reports the value of a response variable with respect to time. Time-weighted statistics have the following characteristics:

- Values persist over time.
- During simulation, the following values are maintained: the current value of the variable, the time of the last change in value, the cumulative value-time product.
- The summary statistic is the time-weighted value of the variable.

An example of a time-weighted statistic would be the average number of entities in the system. In this case, every time an entity enters or leaves the system, the current count is multiplied by the time since the last change. This product is then added to a cumulative, entity-time value. The current count is then updated together with the time since the last change which is updated to the current clock time. A example graph of this time-weighted statistic is shown in Figure 7.1.

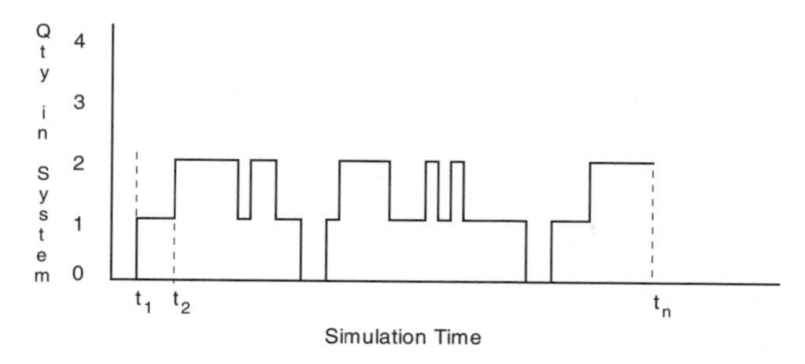

Figure 7.1. Graph Showing Quantity of Entities in the System Over Time.

At the end of the simulation, the average number of entities in the system can then be computed by:

Avg. entities in system = Cumulative entity-time product / Simulation length

If x_i represents the ith observation, t_i represents the duration of the ith observation, and T represents the simulation time, we can express the average number of entities in the system by:

$$\text{Avg. entities in system} = \Sigma(t_i x_i) / T$$

Both observational and time-dependent statistics provide useful information about model behavior. Next we will look at specific measures that are typically reported following a simulation run.

OUTPUT MEASURES

There are many measures of performance in a simulation. Performance measures can typically be reported on all model elements including:

- Entities
- Locations
- Stations
- Queues
- Resources
- Variables
- Attributes

Often, it is desirable to select specific entities, locations, etc. for which you want to have statistics reported. You may even like more detailed statistics on certain key elements of the model. Simulation software is usually able to provide either general or detailed statistics on selected model elements. A list of specific outputs for both general and detailed statistics for each element type is provided below. These listings are not meant to be an exhaustive list, but are only intended to provide a sampling of typical standard output measures provided by many simulation software products.

General Entity Statistics

- Number of entities, by type, to enter each location.
- Number of total entities, by type, to exit the system.
- Number of total entities remaining in the system.
- Average number of entities in the system.
- Average time spent waiting, processing, in-transit.

Detailed Entity Statistics

- Time spent at each location for each entity.
- Time spent in the system for each entity.
- Time spent waiting, processing, in-transit for each entity.

General Queue Statistics

- Total number of entities to enter each queue.
- Average time per entry.
- Average contents.
- Maximum contents.
- Percentage time down.

Detailed Queue Statistics

- Queue contents over time.
- Histogram of waiting times.

General Workstation Statistics

- Total number of entities to enter each station.
- Average time per entry (Unit-hours occupied / Number of entries).
- Utilization (Unit-hours occupied / Unit-hours of capacity).
- Percentage time down.
- Number of setups.
- Average time per setup.
- Percentage of time in each state (operating, idle, blocked, waiting, down, setup, etc.).

Detailed Workstation Statistics

- Contents over time.
- Time each entity spent in the location.
- Gantt chart showing entities at each location over time, by type.

General Resource Statistics

- Number of times used.
- Average time per use.
- Percent utilization.
- Percent down.
- Percent idle.
- Percent traveling between locations.

Detailed Resource Statistics

- Quantity in use over time.
- Utilization by individual unit.

General Variables and Attribute Statistics

- Average value.
- Minimum value.
- Maximum value.
- Final value.

Detailed Variables and Attribute Statistics

- Value changes over time.
- Value histogram.

In addition to these standard output measures, most simulation software also provides the capability for a modeler to gather and report statistics that are not a standard part of the output. This is done through variable tracking and writing to external files. A user is often allowed to customize his or her own reports using whatever format and information provided by the simulation desired.

STATISTICAL ANALYSIS OF SIMULATION OUTPUT

Stochastic simulation follows the random input, random output (RIRO) principle. Since random input variables are used to drive the model (activity times, etc.), then the output measures of the simulation (throughput rate, average waiting times, etc.) are also going to be random. Suppose, for example, we are interested in determining the throughput of a manufacturing system for a given time period. Suppose, further, that the system has a number of random variables such as machining times, time between machine failures, repair times, etc. Running the simulation once provides a random sample of what might happen under such conditions. Changing the initial random seed value and running the simulation again would produce a different result. Clearly, running the simulation once is inadequate to get any kind of precise estimate of the expected behavior of the system.

To draw correct inferences from simulation output, it is essential to have an understanding of basic inferential or descriptive statistics. In descriptive statistics, we deal with a population, a sample, and a sample size. The *population* is the entire set of items or possible outcomes (e.g., all possible rolls of a die). A *sample* would be one, unbiased observation or instance of the population (e.g., a single roll of a die). The *sample size* is the number of samples we are working with (e.g., a particular number of rolls of a die). The idea behind inferential statistics is to gather a large enough sample size to draw valid inferences about the population while keeping the sample gathering time and cost at a minimum.

To conduct an experiment using a simulation model, the simulation is run once to get one replication of the experiment. The outcome of a replication represents a single sample. To obtain a sample size *n*, *n* replications of the experiment need to be run. To obtain a statistically varying sample for each replication, a different starting seed must be used for each replication. Suppose, for example, that we are interested in determining the mean or average expected time (μ) entities spend waiting in a queue. We could make *n* replications of the simulation for periods of identical operating conditions, changing only the initial random seed. This would give us *n* sample waiting time observations ($x_1, x_2, \ldots, x_n$). We could then calculate the average waiting time value ($\overline{X}$) for our sample size *n* as follows:

$$\overline{X} = \frac{\sum_{i=1}^{n} x_i}{n}$$

Suppose, for example, five replications are made with the following waiting times reported for each replication:

Replication	Waiting Time (minutes)
1	44
2	36
3	38
4	47
5	39

If x_i is the waiting time value reported by the *i*th replication, the expected waiting time can be estimated by calculating $\overline{X}$ as follows:

$$\overline{X} = \frac{\sum_{i=1}^{5} x_i}{5}$$

$$= (44 + 36 + 38 + 47 + 39)/5$$

$$= 40.80 \text{ minutes}$$

This provides only an estimate of the true mean value (μ) since there are only a limited number of samples. Taking additional samples would give us a more accurate estimate of μ. We would, however, need to make an infinite number of replications to determine the true mean or expected waiting time.

The standard deviation (σ), which is a measure of the spread of sample values, for the number of entities in the queue can be estimated by calculating a standard deviation (σ) for the sample size as follows:

$$s = \sqrt{\frac{\sum_{i=1}^{n} [X_i - \overline{X}]^2}{n - 1}}$$

For our example in which a sample size of 5 gave a mean value of 40.8 minutes, the standard deviation (s) would be:

$$s = \sqrt{\frac{\sum_{i=1}^{5} [X_i - 40.8]^2}{4}}$$

$$= 4.55$$

$\overline{X}$ and s are referred to as point estimators of a variable. In this case they are the point estimators for the waiting time in the queue.

Confidence Interval Estimation

Having calculated mean and standard deviation values based on a sample size of n runs, we might be interested in knowing how close our sample size average ($\overline{X}$) is to the true mean (μ). The method used to determine this is referred to as *confidence interval estimation.* A confidence interval is a range within which we can have a certain level of confidence that the true mean falls. For a given confidence level or probability (P), say .90 or 90 percent, a confidence interval is calculated by using $\overline{X}$ as the midpoint of the interval and computing an interval half-width (*hw*) which is the portion of the interval which lies on either side of $\overline{X}$. Our interval is then expressed as $\overline{X} \pm hw$. A confidence interval, then, is expressed as the probability (P) that the unknown true mean (μ) lies within the interval $\overline{X} \pm hw$. An explanation of how confidence interval estimates are computed is contained in Appendix C at the end of this book.

Number of Replications

There are methods for determining the number of replications to run to establish a particular confidence level for a specified amount of error between the point estimate of the mean ($\overline{X}$) and the theoretical true mean (μ). First, we let α represent the probability that the error between $\overline{X}$ and μ *will* exceed a specified error amount denoted by the letter *e*. If $\alpha = 0.10$ (there is a 10 percent chance that the difference between $\overline{X}$ and μ will exceed an amount *e*), then we can be (1-α) times 100 percent confident that the difference will not exceed a specified amount *e*.

STATISTICAL PROBLEMS WITH SIMULATION OUTPUT

There are several problems in using traditional statistical methods in analyzing simulation output. We have explained how a single simulation run represents a only a single sample of possible outcomes from the simulation. Further, we have shown that averaging a large number of runs comes closer to the true expected performance of the model, but is still only a statistical estimate. We have also shown how to construct a confidence interval for an expected output. What we have not mentioned is that using standard methods for analyzing simulation output is based on the following three important assumptions about sampled observations:

- Observations are *independent* so that no correlation exists between consecutive observations.
- Observations are *identically distributed* throughout the entire duration of the process (they are time invariant).
- Observations are *normally distributed.*

Unfortunately, the first two assumptions rarely hold true with regard to simulation output. Observations are dependent as is evidenced, for example, by examining waiting times in queues during a simulation. When the simulation starts and the first entity begins processing, there is no waiting time in any queue. Obviously, the more congested the system becomes at various times throughout the simulation, the longer entities will be waiting in queues. If the waiting time observed for one entity is long, it is highly likely that the waiting time for the next entity observed is going to be long and vice versa. Observations exhibiting this correlation between consecutive observations are said to be *autocorrelated.*

In addition to being autocorrelated, output observations from a simulation run are frequently *nonstationary* in that they do not follow the same distribution throughout the simulation run. This is due to the time-varying behavior of many systems in which the statistical distributions used for defining the behavior of the system vary over time. If we look at waiting times for many service systems, the expected waiting time during one part of the day is not likely to be represented by the same distribution as during another part of the day.

The assumption that observations are normally distributed is often violated in simulation. Individual waiting times in a queue, for example, tend to follow a skewed distribution. In order to ensure that output is normally distributed, it is frequently necessary to take a large sample size where each observation is itself an average of a large number of observations. According to the *central limit theorem* of statistics, the distribution of the multiple sample averages tends to be normally distributed. Gathering large samples, of course, can be time consuming if the model is large and the simulation is done on a slow desktop computer.

In order to achieve statistically significant results from a simulation, these three factors must be taken into consideration. A repeated precaution to be made at this point is that in determining the statistical significance of the output of a simulation run or runs, we are talking only about the statistical significance relative to the model itself, not relative to the actual system. So that running sufficient replications to produce a 95 percent confidence level with a 5 percent error means only that there is a probability of 95 percent that the replication results are within plus or minus 5 percent of the true performance *of the model only*. There is absolutely no correlation between the statistical significance of the model results and the validity of the model. Model validity is enhanced only by ensuring that the model data is an accurate representation of the actual system.

STEADY-STATE VERSUS TRANSIENT BEHAVIOR

Determining experimental procedures such as how many replications to make and how to gather the statistics depends on whether we are interested in studying the *transient* periods of a system or the long-term *steady-state* behavior of a system. System behavior is transient if there is no constancy or regularity in the statistical fluctuation of performance variables in the system. The system is in a transitional state between operating from one set of conditions to another. For example, when a system starts up, it makes a transition from having no entity arrivals to some positive arrival rate. Once the system has had time to reflect the impact of this arrival rate, the system is said to have reached a steady state. During steady state, the statistical average behavior of the system does not change over time. Once in a steady state, if the operating rules change or the rate at which entities arrive changes, the system reverts again to a transient state until the system has had time to start reflecting the long-term behavior of the new operating circumstances. A system may be in a transient state for any one of three reasons:

1. The system is starting empty and takes time to reach steady state.
2. System variables are time varying and a transition occurs when going from a steady state for a time period to the steady state of the next time period.

3. A single abnormal disruption to the system (a strike or abnormally long repair time) causes the system to go through a transitional period before regaining steady state.

One example of a system in a transient state is a restaurant in which waiting times tend to increase or decrease depending on the time of day. A manufacturer of consumer products, on the other hand, may have about the same steady-state fluctuation in production throughout the entire day. As a point of clarification, a steady-state condition is not one in which the observations are all the same, nor even that the variation in observations is any less than during a transient condition. It only means that all observations throughout the steady-state period will have approximately the same distribution.

TERMINATING VS. NONTERMINATING (STEADY-STATE) SIMULATIONS

As part of setting up the simulation experiment, one must decide what type of simulation to run. Simulations are usually distinguished as being either *terminating* or *nonterminating* (Law 1991). The difference between the two has to do with whether or not there is an obvious way for determining the length of the simulation. Different measures of interest are also associated with whether the simulation is terminating or nonterminating.

A terminating simulation is one for which the system has a natural, and therefore obvious, event that determines when the simulation should end. A natural event might be the completion of a set of jobs or the closing of a business at the end of a day. Consider, for example, an aerospace manufacturer that receives an order to manufacture 200 airplanes of a particular model. The company may be interested in knowing how long it will take to produce the aircraft along with existing workloads. The simulation run should be terminated when the 200th plane is completed since that covers the period of interest. Another example of a terminating simulation is one that models a condition that exists during only a single period of time. It may be known, for example, that the production schedule for a particular item changes weekly. At the end of each 40-hour cycle, the system is "emptied" and a new production cycle begins. In this situation, a terminating simulation would be run in which the simulation run length would be 40 hours.

Terminating simulations are sometimes in a transient state; continually in a state of adjustment between one set of operating conditions and another. Terminating simulations often repeat a cycle of starting empty, becoming busy for a period of time, and finally emptying again. During transitional periods, the system is seldom in any particular state long enough to qualify as a steady-state condition. Consequently, in terminating simulations, we are generally interested in studying the behavior of the system for one or more particular periods of the transient operation cycle. Examples of terminating simulations might include:

- Simulation of most service institutions (banks, restaurants, etc.).
- Simulation of customer service centers.
- Simulation of job shops.

While terminating simulations are frequently in a transient state, they can be, and often do go through, one or more steady states as well.

A nonterminating or steady-state simulation is one for which there is no natural or obvious event or point in time for terminating the simulation. This does not mean that the simulation never ends, nor does it mean that the system being simulated has no eventual termination. It only means that the simulation could theoretically go on indefinitely without affecting the outcome. For nonterminating simulation, the modeler must determine a suitable length of time to run the model. An example of a nonterminating simulation is a model of a manufacturing operation in which oil filters are produced on a continual basis at the same pace. The operation runs two shifts with an hour break during each shift. Breaks and third shift hours are excluded from the model since work always continues exactly as it left off before the break or end of shift. The length of the simulation is determined by how long it takes to get a steady-state reading of the model behavior. Other examples of nonterminating simulations include:

- Models of most batch flow or flow line manufacturing systems (e.g., appliance manufacturing).
- Models of document processing.
- Models of hospital emergency rooms.

Nonterminating simulations can and often do change operating characteristics after a period of time, but usually only after a long enough period of time has elapsed to establish a steady-state condition. Take, for example, a production system that runs 10,000 units per week for 5 weeks and then increases to 15,000 units per week for the next 10 weeks. The system would have two different steady-state periods.

In practice, one need not be too concerned about whether a particular system ever exhibits transient behavior. The important question to ask is whether the measure of interest is a characteristic or result of steady-state behavior or if it also includes some transient behavior. Noting that the nature of the system itself, whether it be terminating or nonterminating, has little to do with whether the system has any steady-state or transient periods. Hoover and Perry (1990) have observed:

> The difference between the transient and steady-state properties of a system are often misunderstood and confused as being equivalent to a system that is terminating or nonterminating. When simulating a terminating system, it is possible that during most of the simulation, the system is in the steady-state phase. . . . But it is also not unusual to simulate a nonterminating system that has no steady-state phase.

ANALYSIS OF TERMINATING SIMULATIONS

Sample runs for terminating simulations are usually conducted by making multiple short simulation runs of the period of interest using a different random seed for each run. This procedure, called replicating the simulation, enables statistically independent and unbiased observations to be made at designated points in time or for designated periods during each replication. Statistics are also often gathered on overall performance for each replication. Since simulation utilizes a pseudo-random number generator for generating random numbers, running the simulation multiple times simply reproduces the same sample. In order to get an independent sample, the starting seed value for each random stream must be different for each replication, thus ensuring that the random numbers generated from replication to replication are independent.

For terminating simulations, we are generally interested in the changing pattern of behavior over time as much, if not more, than the overall average behavior. It would be absurd, for example, to conclude that because two servers (waiters or waitresses) are only busy an average of 40 percent during the day that only one server is needed. This average measure reveals nothing about the utilization of the servers during peak periods of the day. A more detailed report of waiting times during the entire work day may reveal that three servers are needed to handle peak periods, whereas only one server is necessary during off-peak hours. In this regard, Hoover and Perry (1990) note:

> It is often suggested in the simulation literature that an overall performance be accumulated over the course of each replication of the simulation, ignoring the behavior of the systems at intermediate points in the simulation. We believe this is too simple an approach to collecting statistics when simulating a terminating system. It reminds us of the statistician who had his head in the refrigerator and feet in the oven, commenting that on the average he was quite comfortable.

Two important questions to answer when experimenting with a terminating simulation are:

1. What is the period of interest to be simulated?
2. How many replications should be run?

Often the period of interest is a cycle such as a daily cycle, or, if a pattern occurs over a week's time, the cycle is weekly. Some cycles may vary monthly or even annually. Cycles need not be repeating to be considered a cycle. Some manufacturers, for example, may be interested in the ramp-up period of production during the introduction of a new product which is a one-time occurrence.

ANALYSIS OF STEADY-STATE BEHAVIOR

The key issues in modeling steady-state phenomena are somewhat different than those associated with terminating simulations. In steady-state simulations, the following issues must be dealt with:

- Determining and eliminating the initial warm-up bias.
- Selecting among several alternative ways for obtaining sample observations.
- Determining run length.

Determining the Warm-Up Period

In a steady-state simulation, we are interested in the steady-state behavior of the model. Since a model starts out empty, it usually takes some time before it reaches steady state. In a steady-state condition, the response variables in the system (processing rates, utilizations, etc.) exhibit statistical regularity (the distribution of these variables are approximately the same from one time period to the next). Figure 7.2 illustrates the typical behavior of a response variable Y as the simulation progresses through eight periods of a simulation.

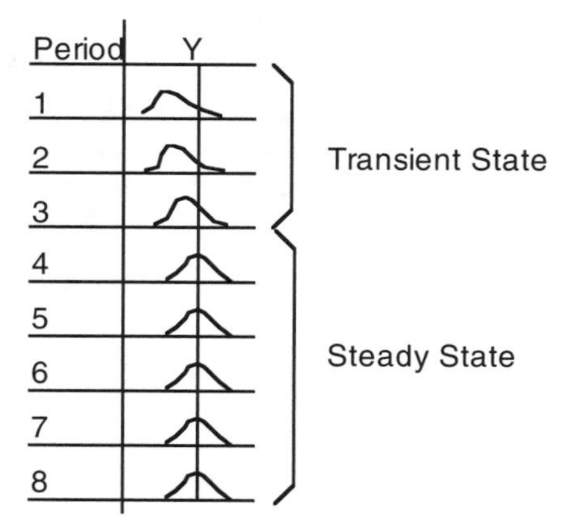

Figure 7.2 Behavior of Response Variable Y for Successive Periods During Simulation.

The time that it takes to reach steady state is a function of the activity times and the amount of activity taking place. For some models, steady state might be reached in a matter of a few minutes of simulation time. For other models it may take several hundred hours to reach steady state. In modeling steady-state behavior, we have the problem of determining when a model reaches steady state. This startup period is usually referred to as the *warm-up period*. We want to wait until after the warm-up period before we start gathering any statistics. This way we eliminate any bias due to observations taken during the transient state of the model.

While several methods have been presented for determining warm-up time (Law and Kelton 1991), the easiest and most straightforward approach, although not necessarily the most reliable, is to run a preliminary simulation of the system, preferably with several (3 to 5) replications, and observe at what time the system reaches statistical stability. The length of each replication should be relatively long and allow even rarely occurring events such as infrequent downtimes to occur at least two or three times. This approach assumes that the mean of the changing response variable distribution is the primary indicator of convergence rather than the variance, which would appear to be often the case. To determine a satisfactory warm-up period using this method, one or more key response variables should be monitored by period over time, such as the average number of entities in a queue or the average utilization of a resource. It is preferable, if possible, to reset the response variable after each period rather than track the cumulative value of the variable, since cumulative plots tend to average out instability in data. Once these variables begin to exhibit steady state, we can add a 20 to 30 percent safety factor and be reasonably safe in using that period as the warm-up period. This approach is simple, conservative, and usually satisfactory. Remember, the danger is in underestimating the warm-up period, not overestimating it. Relatively little time and expense are needed to run the warm-up period longer than actually required. Figure 7.3 illustrates averaging the number of entities processed by hour for several replications. Since statistical stability is reached at about 10 hours,12 to 15 hours is probably a safe warm-up period to use for the simulation.

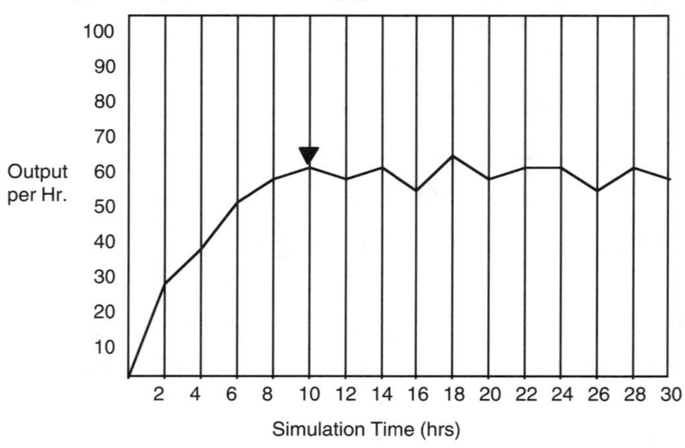

Figure 7.3 Plot of Hourly Entity Output to Identify Start of Steady State.

Obtaining Sample Observations

In a terminating simulation, sample observations are made by simply running multiple replications. For steady-state simulations, we have several options for obtaining sample observations. Two approaches that are widely used are:

1. Running independent replications
2. Interval batching

Running independent replications. This method of obtaining sample observations is very similar to that used for terminating simulations. The only difference is that the initial warm-up bias must be determined and eliminated, and an appropriate run length must be determined. Once the replications are made, confidence intervals can be computed as described earlier in this chapter.

One advantage of running independent replications is that it ensures that samples are independent. On the negative side, running through the warm-up phase for each replication slightly extends the length of time to perform the replications. Furthermore, there is a possibility that the length of the warm-up period is underestimated, causing biased results.

Interval batching. Interval batching (also referred to as the *batch means technique*) is a method in which a single, long run is made with statistics being reset at specified time intervals. This allows statistics to be gathered for each time interval and a mean is calculated for each interval batch. Since each interval is correlated to the previous and the next intervals (called serial correlation or autocorrelation), the batches are not completely independent. The way to gain greater independency is to use large batch sizes and to use the mean values for each batch.

When using interval batching, confidence interval calculations can be performed. The number of batch intervals to create should be at least ten and possibly more depending on the desired confidence interval.

Determining Run Length

Determining run length for terminating simulations is quite simple since there is a terminating event that defines it for us. Determining the run length for a steady-state simulation is more difficult since the simulation could be run indefinitely. The benefit of this, however, is that we can produce good representative samples. Obviously, running extremely long simulations is impractical, so the issue is to determine an appropriate run length to ensure that an adequately representative sample of the steady-state response of the system is taken.

The recommended length of the simulation run for a steady-state simulation is dependent upon the interval between the least frequently occurring event and the type of sampling method (replication or interval batching) used.

If running independent replications, it is usually a good idea to run the simulation enough times to let every type of event (including rare ones) happen at least several times and, if practical, several hundred times. Remember, the longer the model is run, the more confident you can become that the results represent the steady-state behavior. If collecting batch mean observations, it is recommended that run times be as large as possible to include at least 1,000 occurrences of every type of event (Thesen and Travis 1992).

COMPARING ALTERNATIVE SYSTEMS

Simulations are often performed to compare two or more alternative designs. This comparison may be based on one or more decision variables such as buffer capacity, work schedule, or resource availability. Comparing alternative designs requires careful analysis to ensure that the differences that are being observed are attributable to actual differences in performance and not to statistical variation. This is where running multiple replications may again be helpful. Suppose that Method A for deploying resources yields a throughput of 100 entities for a given time period while Method B results in 110 entities for the same time period. Is it valid to conclude that Method B is better than Method A, or might additional replications actually lead the opposite conclusion?

Evaluating alternative configurations or operating policies is properly done by using hypothesis testing. In hypothesis testing, first of all a hypothesis is formulated (that Methods A and B both result in the same throughput). A test is then made to see whether the results of the simulation lead us to reject the hypothesis. The outcome of the simulation runs may cause us to reject the hypothesis of equal throughput capability and conclude that the method used does indeed depend on which method is used.

Sometimes there may be insufficient evidence to reject the stated hypothesis and thus the analysis proves to be inconclusive. This failure to obtain sufficient evidence to reject the hypothesis may be due to the fact that there really is no difference in performance, or it may be a result of the variance in the observed outcomes being too high given the number of replications to be conclusive. At this point, either additional (perhaps time consuming) replications may be run or one of several variance reduction techniques might be employed (see Law and Kelton 1991).

USE OF RANDOM STREAMS

One of the most valuable characteristics of simulation is the ability to both reproduce and randomize replications of a particular model. Simulation allows probabilistic phenomena within a system to be controlled or randomized as desired for conducting controlled experiments.

A stream is a sequence of independently cycling, unique random numbers that are uniformly distributed between 0 and 1 (see Figure 7.4). Random number streams are used to generate additional random numbers from other probability distributions (normal, beta, gamma, etc.). After sequencing through all of the random numbers in the cycle, the cycle starts over again through the same sequence. The length of the cycle before it repeats is called the cycle period, which is usually very long.

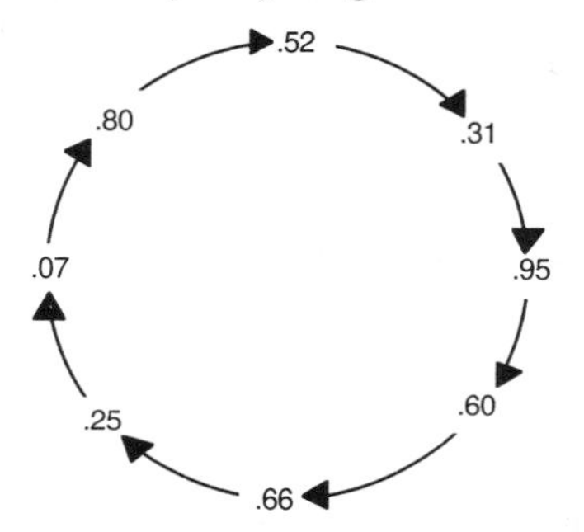

Figure 7.4 Example of a Random Stream Cycle With a Very Short Period.

A random stream is generated using a random number generator or equation. The random number generator begins with an initial seed value. After that each successive value uses the previous value as input to the generator. Each stream used in a simulation has its own independent seed and tracks its own values for subsequent input to the generator. Where the sequence begins in the cycle depends on the initial seed value used by the generator.

Any time a particular number seeds a stream, the same sequence of values will be repeated every time the same seed is used to initialize the stream. This means that various elements within a model can be held constant with respect to their performance while other elements vary freely. Simply specify one random number stream for one set of activities and another random number stream for all other activities.

Because the same seed produces the same sequence of values every time it is used, completely independent functions within a model must have their own streams from the start. For example, arrival distributions should generally have a random number stream that is used nowhere else in the entire model. That way, activities added to a model that sample from a random number stream will not inadvertently alter the arrival pattern because they do not affect the sample values generated from the arrival distribution.

For an example of how multiple streams can be useful, consider two machines, Mach1 and Mach2, which go down approximately every 4 hours for servicing. To model this, the frequency or time between failures is defined by a normal distribution with a mean value of 240 minutes and a standard deviation of 15 minutes. The time to repair is 10 minutes. If no stream is specified for each normal distribution, the same stream is used to generate sample values for both machines. So, if the next two numbers in the stream number are .21837 and .86469, Mach1 gets a sample value from the normal distribution that is different from Mach2. Therefore, the two machines will go down at different times.

Suppose that the resource that services the machines must service them both at the same time, so we would like to have the machines go down at the same time. Using the same stream to determine both downtimes will not bring them down at the same time, because a different random number will be returned from the stream with each call to generate a random normal variate. Using two different streams, each dedicated to each machine's downtime and each having the same initial seed, will ensure that the machines each go down at the same time every time. The two streams have the same starting seed value so they will produce exactly the same sequence of random numbers.

SUMMARY

Correctly interpreting simulation results is essential to making sound decisions based on simulation output. Simulation provides useful information if one knows how to properly use it. While "ballpark" decisions require little analysis, more precise decision making requires more careful analysis and extensive experimentation. Experiments can range from simple, single replication runs to multiple scenario, multiple replication runs.

This chapter described some key considerations involved in analyzing simulation output. Differences between terminating and nonterminating simulations were explained. Practical guidelines were provided for determining run length, number of replications, and warm-up period. We also discussed issues for consideration when comparing alternative systems. The importance of random number streams and seeds were explained with examples.

Chapter 8
Modeling Manufacturing Systems

"Production is not the application of tools to materials, but rather the application of logic to work."

Peter Drucker

INTRODUCTION

Manufacturing systems are processing systems in which raw materials are transformed into finished products through a series of operations performed at workstations. In the rush to get new manufacturing systems online, engineers and planners often become over preoccupied with the processes, equipment, and methods without fully considering overall coordination, integration, and scheduling issues. Manufacturing systems wind up being implemented that are poorly conceived, and often fall below anticipated levels of performance.

In Chapter 6, we discussed general procedures for modeling elements common to both manufacturing and service systems. This chapter discusses design and operating considerations that are more specific to manufacturing systems, and how to take these considerations into account in a simulation model. Different types of manufacturing systems are presented together with their respective performance measures, decision variables, and method of representation in a simulation model.

APPLICATIONS OF SIMULATION IN MANUFACTURING

To achieve optimum performance in a manufacturing system, it is important to make intelligent design and management decisions. Simulation has proven to be an effective tool in helping to sort through the complex issues surrounding manufacturing decisions. As companies move towards greater integration and implementation of total manufacturing resource planning systems, simulation is recognized as an essential tool to effectively plan and control resources (Greene 1987). Simulation helps evaluate the performance of alternative designs and the effectiveness of alternative production plans. The decision or experimental variables that simulation addresses may be categorized as being either design or management related. A list of typical issues or questions to be answered under each category is provided below:

I. System Design Decisions

- What type and quantity of machines or workstations should be used?
- What type and quantity of auxiliary equipment and operators are needed?
- How much tooling and fixturing is required?
- What is the production capability (throughput rate) of a given system?
- What type and size of material handling system should be used?
- What are the optimum number and size of storage areas and buffers?
- What is the best layout of workstations?
- What is the most effective control logic?
- What is the optimum unit load size?
- How effectively are resources being utilized?
- What effect does a process or method change have on overall production?
- How balanced is the work flow?
- Where are the bottlenecks?
- What is the impact of machine downtime on production (reliability analysis)?
- What is the effect of setup time on production?
- What is the effect of centralized versus localized storage?
- What is the effect of vehicle or conveyor speed on part flow?
- How many repair personnel are needed?
- What is the overall effect of automating an operation?

II. System Management Decisions

- What is the best way to schedule preventive maintenance?
- How many shifts are required to meet production requirements?
- What is the optimum production batch size?
- What is the optimum sequence for producing a set of jobs?
- What is the best priority rule for selecting a particular job?
- What is the best way to allocate resources for a particular set of tasks?
- What is the effect of a preventive maintenance policy as opposed to a corrective maintenance policy?
- How much time do jobs spend in a system (makespan or throughput time analysis)?

- What is the effect of different product mix combinations?
- How well can a particular production schedule be met?
- What impact would a given inventory policy (just-in-time) have on production?

In addition to being used for making specific decisions, simulation has been used to derive general design rules so that the most cost effective decisions can be made for a range of operating conditions. One study, for example, used simulation to estimate how the output capacity of a manufacturing system changes as the various components of the system are increased or improved for a known cost. This data was then used to develop "cost versus output" curves for a given part or product mix. By comparing these curves for different types of production systems, it was possible to identify break-even points where a particular type of system was economically preferred over other systems for particular products or product mixes (Renbold and Dillman 1986).

DECISION HORIZON FOR MANUFACTURING

The potential benefits of simulation in designing and managing manufacturing systems are many and encompass not only a broad spectrum of applications but also span a wide range of planning horizons. Simulation has been identified as an effective decision support tool for long-term, medium-term and short-term planning and control (Suri and Whitney 1984). Figure 8.1 illustrates the potential planning horizons for simulation in manufacturing.

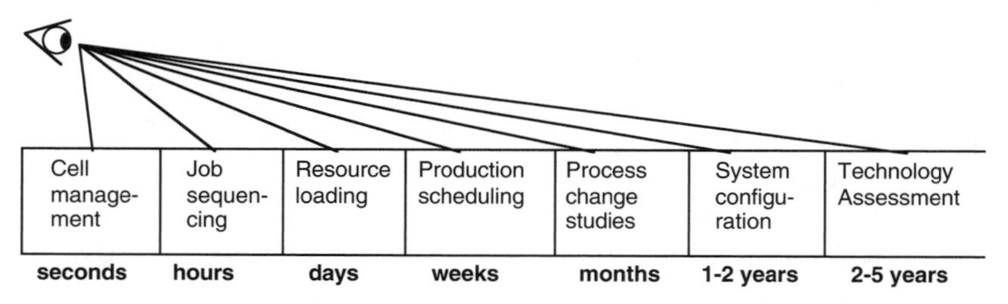

Figure 8.1 Potential Planning Horizons For Using Simulation.

Cell management. During actual production, simulation has been integrated with manufacturing cells to perform real-time decisions such as the next task for a resource to complete or the machine to select for an operation to be performed. These management decisions use the same logic that was used in the initial planning of the system. The benefits of using the simulator itself to manage the system are as follows:

- Logic used in the simulator does not need to be recoded for the controller.

- Animation capabilities allow processes to be graphically monitored.
- Statistical gathering capabilities of simulation automatically provide statistics on selected measures of performance.

Job sequencing. Job sequencing determines which job is performed next on a particular machine. The goal might be to minimize lateness or to maximize utilization. Some of the more common job sequencing rules are presented in the discussion on job shops.

Resource loading. Resource loading determines how resources will be used to implement a particular schedule. It is the capacity planning part of scheduling to ensure that sufficient capacity is available to produce the desired quantity.

Production scheduling. The use of simulation to perform scheduling is called simulation-based scheduling. Scheduling determines the start and finish times for jobs to be produced and is the result of sequencing decisions. Sequencing decisions determine dynamically the order in which jobs are to be run on a particular machine. This is the type of decision that simulation can help make and may be based on nearly any criteria defined by the modeler. Once the simulation is run, a report can be generated showing the start and finish times of each job on each machine. Simulation does not establish a schedule but rather reports the schedule resulting from a given set of operating rules and conditions.

Process change studies. A manufacturing or process engineer who is responsible for ongoing improvement of existing processes can use simulation to evaluate the potential impact of a change on overall production before recommending the proposed change.

System configuration. Simulation is an excellent tool, and perhaps most widely used, to configure a system. System configuration consists of deciding what resources and technologies to use, and how many resources are needed to provide the capacity required.

Technology assessment. For long-range planning, simulation can aid in assessing new technologies that might be useful in the future. This function is often performed in larger corporations by advanced manufacturing groups responsible for strategic planning. Simulation can be used to assess the feasibility of new technologies prior to committing capital funds and corporate resources.

EMULATION

A special use of simulation in manufacturing, particularly in automated systems, has been in the area of hardware emulation. As an emulator, simulation takes inputs from the actual control system (programmable controllers, microcomputers, etc.), mimics the behavior that would take place in the actual system, and then provides feedback signals to the control system. Essentially, the control system is plugged into the model instead of the actual system. In this way, simulation is used to test, debug, and even refine the actual control system before any hardware has been installed. This has helped reduce significantly the time to start up new systems and implement changes to automation. Emulation software has been developed by control system suppliers as well as third-party developers. The first commercially available emulation product was developed by HEI Corporation, which received a patent for the first emulation system.

MANUFACTURING TERMINOLOGY

Manufacturing systems share many common elements and objectives. They also share a vernacular that is often quite foreign to those unfamiliar with manufacturing activities. Much of the discussion on manufacturing systems requires a knowledge of the common terminology used in manufacturing industries. Some of the terminology that will be used to discuss manufacturing systems is given below:

Operation. An operation is the activity performed on a product at a workstation. Operations generally tranform the product in some way, although non-transformational activities such as inspection or test may appropriately be termed operations.

Workstation. A workstation or work center is the place or location where an operation is performed. A workstation is defined in the APICS dictionary as "a specific production facility consisting of one or more people and/or machines which can be considered as a single unit for purposes of capacity requirements planning and detailed scheduling."

NC machine. An NC (numerically controlled) machine is a machine tool whose spindle and table action are controlled by a computer.

Machining center. A machining center is essentially an NC machine with a tool exchanger which provides greater capability than a stand-alone NC machine. Machining centers have been defined as "multifunctional, numerically controlled machine tools, with automatic tool changing capabilities and rotating cutting tools that permit the programmed, automated production of a wide variety of parts" (Wick 1987).

Process. In manufacturing, a process defines: (1) the nature of the operation performed on a part at a workstation (material removal, metal forming) or (2) the entire sequence of operations performed on a part or subassembly.

Routing. The product routing defines the material flow sequence from one workstation to another.

Batch. A batch is a collection or grouping of parts for a particular purpose. There are three general uses of the term *batch* in manufacturing:

1. The production batch or batch run refers to the number of items of the same part or product type that are run consecutively before introducing a new part or product type.
2. The move or transfer batch is the quantity of parts that are moved between workstations.
3. The process batch is the quantity of parts that are processed together at a workstation.

Process plan. The process plan specifies the sequence of operations required to convert raw materials into finished products or subassemblies.

Master schedule. The master schedule defines what end-products are to be produced for a given period of time.

Production plan. The production plan is a detailed schedule of production for each individual component comprising each end-product.

Bottleneck. The bottleneck is the workstation that has the highest utilization (the largest ratio of required processing time to time available). A bottleneck may be any constraint (an operator, a conveyor system) in the production of goods.

Changeover. Changeover is the change from one product or part type to another.

Setup. Setup is the activity required to prepare a resource when a changeover occurs.

Job. Depending on the industry and nature of production, a job may be any of the following:

- In a general sense, a job refers to the activity or task being performed.
- In mass production, a job refers to each individual piece part.
- In a job shop, a job refers to an order to produce a particular quantity of a particular part or product.

Material handling. The movement of parts, tooling and scrap from one location to another.

Cycle time. Cycle time is the time for a machine to perform a single operation. If load and unload times are included, it is referred to as floor-to-floor time.

Resource capacity. Capacity refers to either holding capacity as in the case of a tank or storage bin, or production capacity as in the case of a machine. Production capacity is of the following four types:

- *Theoretical capacity*. What a machine is capable of producing if it was in constant production assuming 100 percent efficiency (the theoretical capacity is simply the period of time of interest divided by the cycle time of the machine).
- *Effective capacity*. The theoretical capacity factored by the reliability of the machine and allowances such as load, unload, and chip removal.
- *Expected capacity*. The effective capacity of the machine factored by anticipated allowances for system influences such as shortages and blockages.
- *Actual capacity*. What the machine produces as a result of actual operation and system interaction.

Scrap rate. Scrap rate is the percentage of defective parts that are removed from the system following an operation.

Reliability. Reliability is usually measured in terms of mean time between failures (MTBF) which is the average time a machine or piece of equipment is in operation before it fails.

Maintainability. Maintainability is usually measured in terms of mean time to repair (MTTR) and is the average time required to repair a machine or piece of equipment whenever it fails.

Availability. Availability is the percentage of total scheduled available time that a resource is actually available for production. Availability is a function of reliability and maintainability.

Preventive or scheduled maintenance. Preventive or scheduled maintenance is periodic maintenance (lubrication, cleaning, etc.) performed on equipment to keep it in running condition.

Unit load. A unit load is a consolidated group of parts that is containerized or palletized for movement through the system. The idea is to minimize handling through consolidation, and provide a standardized pallet, or container as the movement item.

PERFORMANCE MEASURES

Manufacturing systems can have many production performance measures by which to gauge the effectiveness and efficiency of the system. Typical performance measures include the following:

Cost. Cost includes both investment cost and operating cost. Investment cost consists of capital equipment costs and other initial expenses (training, etc.). Operating costs are usually figured on an annual basis and include ongoing direct and indirect labor and material costs.

Throughput capacity. Throughput is the maximum possible output rate (parts per hour) that can be achieved by the system. If the system processes more than one entity type, it becomes more difficult to estimate capacity.

Manufacturing lead time (MLT). Manufacturing lead time or throughput time is the time required to process the product through the manufacturing system (average time in the system). A related measure is the total lead time which is the time from when raw materials are ordered to the time the finished product is delivered. It may be thought of as customer turnaround time. Manufacturing lead time is the sum of the following:

- Setup or changeover time
- Processing time
- Inspection time
- Move time
- Queue or waiting time

Minimizing manufacturing lead time reduces work-in-process inventory. Other names for manufacturing lead time include *flow time*, *system cycle time* (not to be confused with machine cycle time) and *throughput time*.

Processing time. The time during which actual value is being added to the product (value-added time).

Queue or waiting time. The time parts spend waiting for the next location (blocked time), waiting for processing resources (operators) or material handling resources, waiting for parts to become available (an assembly operation), or waiting for some other condition to be satisfied before processing can take place. Waiting time is considered non-value-added time and should be minimized.

Waste. In contrast to value-added activities, waste refers to anything that incurs a cost, yet provides no value to the product. Suzaki (1987) identifies seven types of waste in manufacturing: (1) waste from overproduction, (2) waste of waiting time, (3) transportation waste, (4) processing waste, (5) inventory waste, (6) waste of motion, and (7) waste from production defects. In the broadest sense, anything other than the minimum amount of equipment, materials, space, information, people, and time essential to add value to the business is a non-value-adding activity.

Makespan. Closely related to manufacturing lead time, makespan is defined as the time to process a given set of jobs. Minimizing makespan aims to achieve high utilization of equipment and resources by getting all jobs out quickly.

WIP. WIP or work-in-process defines the level of in-process inventories which may vary from place to place and from time to time. The relationship between throughput or production rate (λ), WIP (L) and manufacturing lead time (W) is given by Little's law (Little 1961).

$$L = \lambda W$$

Utilization. Utilization is the percentage of time that a resource is in productive use as a percentage of total scheduled time. High resource utilization is an indication that resources are being put to use (not necessarily that productivity is high). High resource utilization may also cause delays that create large queues and high manufacturing lead times.

Flexibility. Flexibility is the ability to handle variations in production. Variations may be one or both of the following:

- *Variety*. Change of product type and/or mix.
- *Volume*. Fluctuations in production quantities.

A flexible system is able to respond to changes in product mix and volume without the creation of excessive queues.

Customer responsiveness. Responsiveness is the ability of the system to deliver products in a timely fashion to minimize customer waiting time. In minimizing job lateness, it may be desirable to minimize the overall late time, minimize the number of jobs that are late, or minimize the maximum tardiness.

Yield. From a production standpoint, yield is the percentage of products completed that conform to product specifications as a percentage of the total number of products that entered the system as raw materials.

MODELING CONSIDERATIONS

Manufacturing systems represent a class of processing systems in which entities (raw materials, components, subassemblies, pallet or container loads, etc.) are routed through a series of workstations, queues and storage areas. As opposed to service systems where the entities are often humans, entities in manufacturing systems are inanimate objects with no wills of their own. Entity production is frequently performed according to schedules to fill predefined quotas. In other instances, production is triggered by low finished goods inventories (make-to-stock production) or by customer orders (make/assemble-to-order production).

Manufacturing systems utilize varying degrees of mechanization and automation, both in entity processing and material movement. In the most highly automated facilities, there may be very little, if any, human involvement to be considered in the simulation study. These characteristics have the following implications for modeling manufacturing systems:

- Often operation times have little, if any, variability.
- Entities are introduced to the system at set times or conditions.
- The routing of entities is usually fixed from the start.
- Entities are often processed in batch.
- Equipment reliability is often a key factor.
- Material handling is often an important part of the model.
- Systems often exhibit steady-state behavior.

Of course, the actual modeling requirements will vary according to the type of system being modeled and the purpose of the simulation.

The next section discusses manufacturing system types and relevant issues for simulation.

MANUFACTURING DECISION VARIABLES

Given the complexity and diverse characteristics of manufacturing systems, there are many different decisions to be dealt with in achieving an optimum design and management strategy. Below we discuss some of the major types of decisions and their potential impact on system performance measures.

Layout Decisions

Layout decisions affect the flow of product through the facility. A good layout results in a streamlined flow with minimum movement and thus minimizes material handling and storage costs. Layouts tend to evolve haphazardly over time with little thought given to work flow. This results in non-uniform, multi-directional flow patterns that cause numerous management and material handling problems. Simulation helps identify inefficient work flow patterns and create better layouts to economize the flow of material.

In planning the layout of a manufacturing system, it is important to have a total system perspective of material flow. The overall part flow within a manufacturing facility is generally referred to as the material flow system and begins from the reception of material at the receiving dock and extends to the shipping of material from the facility (see Figure 8.2). These activities often overlap and frequently share common resources so that it is difficult to analyze one activity in complete isolation of the other two.

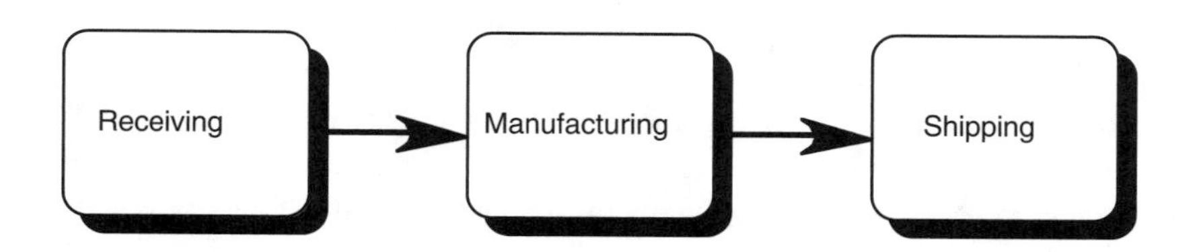

Figure 8.2 Material Flow System.

In laying out equipment in a facility, the designer must choose between a process layout, a product layout, or a part family layout of machines. In a *process layout,* machines are arranged together based on the process they perform. Machines performing like-processes are grouped together.

In a *product layout,* machines are arranged according to the sequence of operations that are performed on a product. This results in a serial flow of the product from machine to machine.

In a *part family layout,* machines are arranged to process parts of the same family. Part family layouts are also called group technology cells since group technology is used to group parts into families.

Figure 8.3 illustrates these three different types of layouts and their implications on material flow.

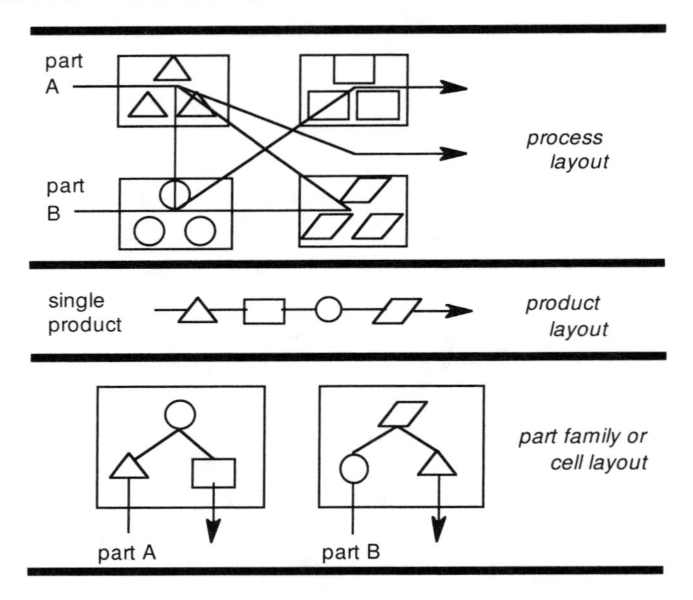

Figure 8.3 Comparison Between a Process Layout, Product Layout, and Part Family or Cell Layout.

Deciding on which layout to use will depend largely on the variety and lot size of parts that are produced. If a wide variety of parts are produced in small lot sizes, a process layout may be the most appropriate. If the variety is similar enough in processing requirements to allow grouping into part families, a cell layout is best. If the variety is small enough and the volume is sufficiently large, a product layout is best. In general, the more the topology resembles a product layout, the greater the efficiencies that can be achieved.

Automation Decisions

One of the decisions during the design phase of a manufacturing system pertains to the type and level of automation. Simulation has been especially helpful in designing automated systems to safeguard against suboptimal performance. As noted by Glenney and Mackulak (1985), "Computer simulation will permit evaluation of new automation alternatives and eliminate the problem of building a six-lane highway that terminates onto a single lane dirt road." Often an unbalanced design occurs when the overall performance is overlooked because of preoccupation with a single element. "Islands of automation" that are not properly integrated and balanced lead to disappointing performance. For example, an automated machine capable of producing 200 parts an hour fed by a manual handler at the rate of 50 an hour will only produce 50 an hour. The expression that "a chain is no stronger than its weakest link" has particular application to automated manufacturing systems.

There are three types of automation to consider in manufacturing: operation automation, material handling automation and information automation. The degree of automation ranges from manual to fully automatic. Figure 8.4. shows the types and degrees of automation.

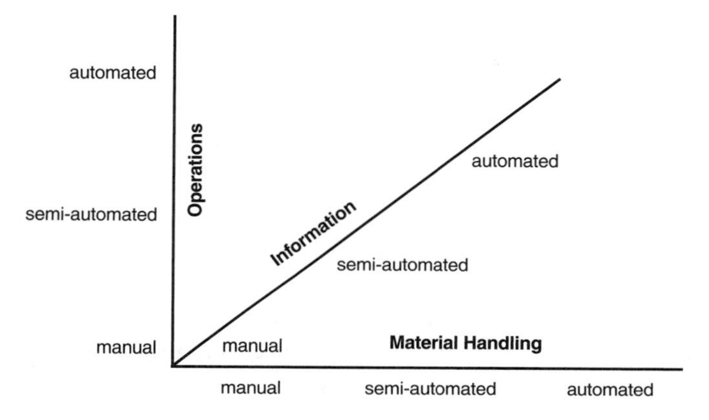

Figure 8.4 Axes of Automation.
(Adapted from Boehlert and Trybula1984)

While information flow activity is frequently overlooked in manufacturing simulations, it may be a contributing factor to the overall system performance. As noted in *Manufacturing Engineering:* "It's easier to make the product than to adjust the paper and data flow required to track the product"(1993).

Automation itself does not solve performance problems. If automation is unsuited for the application, poorly designed, or improperly implemented and integrated, then the system is not likely to succeed. The best approach to automation is to first simplify, then systematize and finally automate. If simplification and systematization cannot be achieved, automation is likely to result in an automated mess. The best application of automation is in situations where activities are well defined and repetitive.

Because simulation can consider multiple variables in manufacturing systems, it is an indispensable tool for achieving the level of system integration envisioned by CIM (computer-integrated manufacturing):

> Integration is the key to any CIM implementation. Computer simulation is a comprehensive analysis tool which allows one to focus on individual components of a manufacturing company without overlooking the integration issues. Thus simulation can be an especially effective tool when planning for CIM (Miner 1987).

Batching Decisions

A batch or lot of parts refers to a quantity of parts grouped together for some purpose. There are typically three different batch types that are often spoken of in manufacturing.

1. Production batch
2. Move batch
3. Process batch

The *production batch* consists of all of the parts of one type that begin production before a new part type is introduced to the system. The *move* or *transfer batch* is the collection of parts that are grouped and moved together from one workstation to another. A production batch is often broken down into smaller move batches. This practice is called "lot splitting." The move batch need not be constant from location to location. In some batch manufacturing systems, for example, a technique called "overlapped production" is used to minimize machine idle time and reduce work in process. In overlapped production, a move batch arrives at a workstation where parts are individually processed. Then instead of accumulating the entire batch before moving on, parts are sent on individually or in smaller quantities to prevent the next workstation from being idle while waiting for the entire batch. The *process batch* is the quantity of parts that are processed simultaneously at a particular operation and usually consists of a single part. The relationship between these batch types is illustrated in Figure 8.5.

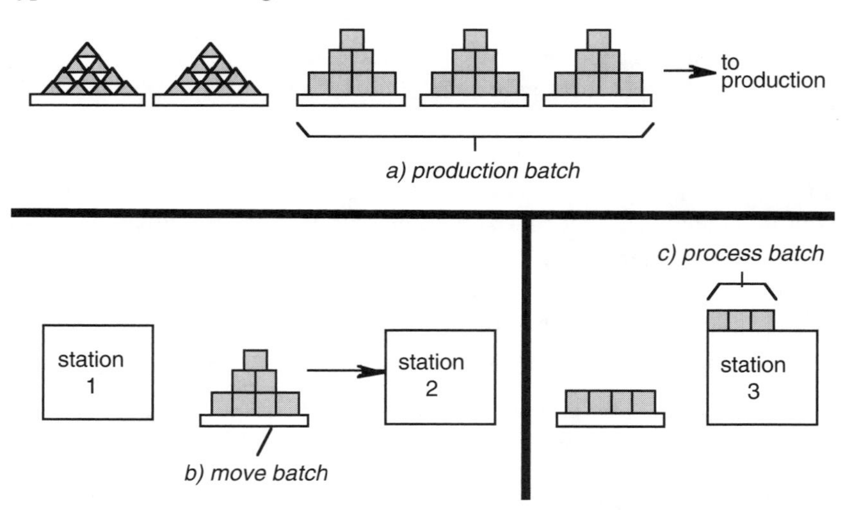

Figure 8.5 Illustration of Production Batch, Move Batch, and Process Batch.

Deciding which size to use for each particular batch type is usually based on economic trade-offs between in-process inventory costs and cost factors that would be reduced by keeping the batch size large (setup costs, handling costs, processing costs, etc.).

Production Control Decisions

Production control determines when and in what quantities parts get processed at individual workstations. It controls the flow of material. On one extreme is a *push system* in which production is driven by workstation capacity and material availability. Each workstation seeks to produce as much as it can, pushing finished work forward to the next workstation. On the other extreme is a *pull system* in which downstream demand triggers each preceding station to produce a part with no more than one or two parts at a station at any given time (see Figure 8.6).

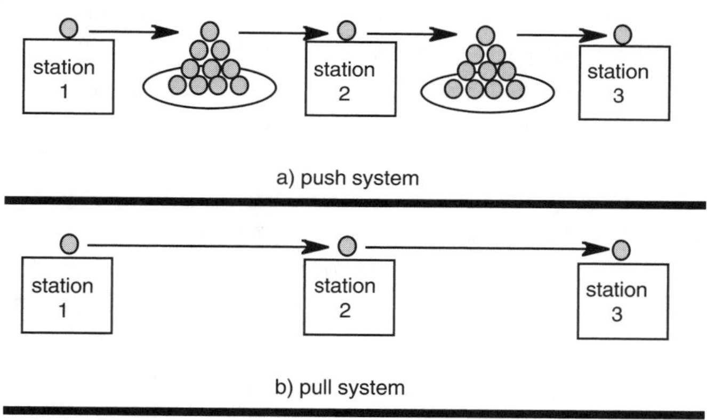

Figure 8.6 Push versus Pull System.

One form of inventory pushing utilizes material requirements planning (MRP) which determines how much each station should produce for a given period. Unfortunately, once planned, MRP is not designed to respond to disruptions and breakdowns that occur for that period. Consequently, stations continue to produce inventory as planned, regardless of whether downstream stations can absorb the inventory.

Pull systems are often associated with the just-in-time (JIT) philosophy which advocates the reduction of inventories to a minimum. The basic principle of JIT is to continuously reduce scrap, lot sizes, inventory, and throughput time as well as eliminate the waste associated with non-value added activities such as material handling, storage, machine setup, and rework. JIT utilizes a demand-pull production control concept sometimes implemented using a method called *kanban* (translated *ticket*). When an operation is completed at the final location and exits the location, a new entity or part is pulled from the preceding WIP buffer. A kanban card accompanies each part bin in the buffer and is returned to the preceding station when the bin moves on. The returned card signals the preceding operation to pull a unit from its own WIP and replenish the kanban.

For most JIT systems where flow is serial, a kanban system is easily modeled by simply limiting the capacity of the workstation buffers to one or two depending on whether the system is a single bin, two bin, etc., kanban system. The term *kanban* is starting to get pushed into the background in favor of the more universally descriptive term *queue-limiter*. This is, after all, what kanban does—it limits queue capacities.

> Kanban is a queue limiter, that is, it tightly links a provider's service with a user's need. Thus, it limits the queue of work in front of the provider (the quantity/amount of items waiting and the time each item waits). Queue limitation replaces lax flow control with discipline, which results in customer satisfaction and competitive advantage (Schonberger and Knod 1994).

Since most simulation languages use capacity limitations to control the flow of entities, entities will not advance to the next station until there is available capacity at that location. This effectively provides a built-in pull system in which the emptying of a buffer or workstation signals the next waiting entity to advance (push systems actually work the same way only buffers are given infinite capacity). Several papers have been published showing the use of limited buffers to model JIT systems (Wang and Wang 1990, Mitra and Mitrani 1990). For more complex pull systems, triggers may be required to signal the release of WIP further than one station upstream.

Inventory Control Decisions

The goal of inventory control is to:

- Maintain sufficient inventory on hand so that shortages do not occur.
- Order enough inventory each time so that the costs associated with frequent ordering are avoided.
- Minimize the inventory carrying costs associated with maintaining unnecessary inventories.

Decisions affecting inventory levels and costs include:

- How much inventory to order when inventory is replenished (economic order quantity).
- What is the point at which an order for more inventory is made (reorder point).
- How much safety stock is desired to have on hand.

There are two primary approaches to deciding when to order more inventory and how much:

1. Reorder point system
2. MRP system

A *reorder point system* is an inventory control system where a particular item is replenished to a certain level whenever if falls below a defined level. MRP, on the other hand, is a bit more complicated in that it is time phased. This means that inventory is ordered to match the timing of the demand for products over the next planning horizon.

Inventory management must deal with random depletion or usage times and random lead times, both of which are easily handled in simulation. To model an inventory control system, one need only define the usage pattern, the demand pattern and the usage policies (priorities of certain demands, batch or incremental commitment of inventories, etc.). The demand rate can be modeled by allowing entities to exit at intervals. Replenishment of raw materials should be made using internally initiated ordering of entities.

Running the simulation can reveal the detailed rises and drops in inventory levels as well as inventory summary statistics (average levels, total stockout time). The modeler can experiment with different replenishment strategies, and usage policies until the best plan can be found that meets the established criteria.

Using simulation modeling over traditional, analytic modeling for inventory planning provides several benefits (Browne 1994):

- *Greater accuracy*. Actual, observed demand patterns and irregular replenishment of inventories can be modeled.
- *Greater flexibility*. The model can be tailored to fit the situation rather than forcing the situation to fit the limitations of an existing model.
- *Easier to model.* Complex formulas that attempt to capture the entire problem are replaced with simple arithmetic expressions describing basic cause-and-effect relationships.
- *Easier to understand.* Demand patterns and usage conditions are more descriptive of how the inventory control system actually works. Results reflect information similar to what would be observed from operating the actual inventory control system.
- *More informative output.* It shows the dynamics of inventory conditions over time and provides a summary of supply, demand, and shortages.
- *More suitable for management.* It provides "what if" capability so alternative scenarios can be evaluated and compared. Charts and graphs are produced that management can readily understand.

Another inventory control issue has to do with the positioning of inventory. Traditionally, inventory was placed in centralized locations for better control. Unfortunately, this created excessive handling of material and increased response times. More recent trends are towards decentralized, point-of-use storage where inventory is kept at the point it is needed. This eliminates needless handling and dramatically reduces response time.

Resource Decisions

Resources used in manufacturing have an obvious impact on overall performance. Besides cost and process capability considerations, the following factors should taken into account when making resource selections:

- Cycle time
- Capacity
- Flexibility
- Reliability
- Maintainability

The cycle time affects the speed at which an operation occurs. If the resource is highly utilized, speed may be a crucial factor in overall system performance. Capacity determines how many parts can be processed or handled at once. Increasing resource capacity is a good way to compensate for other inefficiencies in the system. Flexibility of processing resources can permit greater resource utilization. Material handling resource flexibility permits greater flexibility in routings. Reliability and maintainability together determine the overall availability of the resource.

Capacity Planning Decisions

Capacity planning deals with juggling the workload with capacity to meet planned schedules. Decisions affecting capacity involve the following questions:

- Which resources should work be assigned to?
- How many resources should be used?
- How much time should each resource be scheduled for?

Production Scheduling Decisions

In production scheduling, decisions must be made as to when and in what quantities items should be produced. Traditional scheduling methods fail to account for fluctuations in quantities and resource capacity. In utilizing simulation for production scheduling, short time periods are used, and therefore there is less concern for long-term statistical fluctuations in the system, such as machine downtimes. These points distinguish simulation-based scheduling:

- Model must capture an initial state.
- Usually based on expected times.
- Anomaly conditions are ignored.
- The simulation is run only until the required production has been met.

TYPES OF MANUFACTURING SYSTEMS

Modeling of manufacturing systems requires an understanding of the types of manufacturing systems that exist and the objectives and issues associated with each type of system. The type of manufacturing system is dictated largely by the manufacturing objectives as well as by product volume and variety. There are six types of systems that are common to most discrete part manufacturing:

1. Project shop
2. Job shop
3. Cellular manufacturing
4. Flexible manufacturing systems
5. Batch flow shop
6. Line flow systems (production/assembly lines, transfer lines)

This classification of manufacturing system types is not intended to imply that all manufacturing systems fall neatly into one of these six types. On the contrary, many systems have characteristics of more than one type. For example, some facilities, called mixed shops, combine both functional and linear equipment structures at different stages in the manufacturing process. The suitability of each system type is related largely to the volume and variety of the product produced. Figure 8.7 shows the typical range of suitability for each production system type.

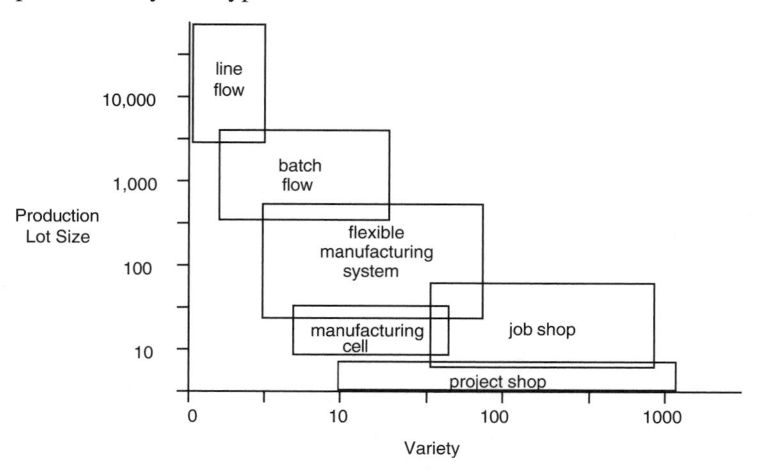

Figure 8.7 Typical Volume and Variety for Each Manufacturing System Type.

A major distinction between each type of system from a material flow perspective is the move quantity and routing. Table 8.1 shows the different flow characteristics of each system type.

Table 8.1 Flow Characteristics of Manufacturing Systems

System	Move Quantity	Routing
Project shop	None	None
Job shop	Large	Random
Cellular	Small	Linear
FMS	Small	Random
Batch flow	Large	Linear
Line flow	Individual	Linear

A description of each of these six major types of manufacturing systems is presented together with the performance measures, decision variables, and modeling representations.

PROJECT SHOP

In a project shop, the part or batch of parts is stationary while the resources are brought to the product for processing. The product is often quite large, such as a building or a ship, and requires considerable time to complete. Production quantities are usually low with products often being produced from start to finish one at a time. Projects typically involve many resources where the quality of the end-product and the time to finish the product are dependent on the capability of the resources as well as the planning and scheduling of those resources. Often, multiple overlapping projects must be coordinated, all requiring the use of common resources. A project is not a true processing system. A project is a progression through time rather than the progression of entities through space (see Figure 8.8).

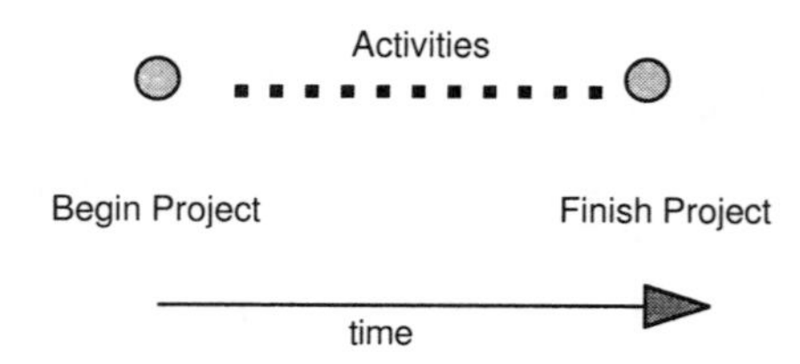

Figure 8.8 Progression of a Project Through Time.

Because of the time and cost of the project, the quality of the end-product is of utmost importance. Delivery time is also a key consideration and penalties are sometimes assessed if due dates are not met. Project cost, although important, is less of a concern. Customer feedback is frequent and can have an impact on the priority and scheduling of remaining activities. Customer commitment is very high because it is difficult to find another producer or to start from scratch once the project is under way.

Projects are typically modeled using CPM (Critical Path Method), PERT (Program Evaluation and Review Technique), or Gantt charting. Many project management software packages have become available to help organize and track the progress of projects. Simulation may be useful, however, if there is significant variation in individual activity times or if shared resources are used that create multiple interdependencies.

Examples

- Software development
- Building construction
- Shipbuilding
- Aircraft manufacturing
- Satellite construction
- Special machine tool manufacturing.

Performance Measures

- Estimated time to complete the project.
- Estimated time to complete multiple projects.
- Probability of completing the project by a particular time.
- Estimated cost to complete the project.
- Expected utilization of resources.
- Potential delays along the critical path due to resource unavailability.

Decision Variables

- The subdivision of the project into individual tasks.
- The time required to complete each task.
- The resources required to perform each task.
- The priorities with which activities access resources.
- The order in which multiple projects are performed.

Questions to be Answered

- What is the optimum project plan that minimizes total costs?
- What is the length and/or cost of a project given a defined set of activities?
- What is the best use of resources to minimize the delay of a project?
- How many resources are needed to meet a particular deadline?
- What is the best coordination of multiple projects to minimize delays?

Model Representation

With a process-oriented simulation language, modeling a project is not as straightforward as modeling process flows. The following table describes at least one way to develop a project model using process-oriented simulation.

System Element	***Model Representation***
Projects	Each project being studied should be represented by a separate entity. Parts of a project that may be performed in parallel or independently of one another should be represented by separate entities. If multiple items or projects are performed separately, then a separate entity should be used.
Activities	The activities can be represented by locations, with each location representing each activity or stage of the project. In modeling a project, we are not so interested in the physical space that the project occupies, but rather in the time phase in which each activity is performed. Activity times may vary in duration.
Resources	Resources can be modeled using the resource construct available in most simulation languages. Resources represent the workers, equipment, etc., required to perform the project. Often multiple resources are required to perform each activity. Priorities are important in a project and some activities may even preempt other activities. Scheduling of resources, shifts, overtime, and learning curves (expertise) are all important.
Activity Sequence	The activity sequence is modeled as the routing for the entity. The routing may specify activities that are performed either in parallel or in series.

Simulation Procedure

Once a model has been built, it is usually necessary to run multiple replications of the model. Since activity times are generally variable, a single simulation run produces only a single observation or possible outcome. Multiple replications will produce several observations which will give a more accurate estimate and confidence interval for the time required to complete the project.

JOB SHOP

A job shop consists of an arrangement of workstations, usually by function or by the type of process they perform. Job shops generally produce a wide variety of parts in low volume. Based on the operations required, each part type has a routing that may not be defined until the actual production time. This dynamic routing allows machines to be selected based on availability since some operations can often be performed on more than one machine. Job shops are still the most common type of manufacturing: it is estimated that 75 percent of all manufacturing is done in production batch sizes of 50 items or less. Job shops may be either a "one off" type in which only one of a kind is produced (tooling or prototype shop) or a repeating type in which quantities are usually greater than one and similar jobs are produced again in the future. In the case of job shops, the general range of processes anticipated are considered in the selection of equipment. General purpose equipment is used that is capable of providing processes for a broad range of products (see Figure 8.9).

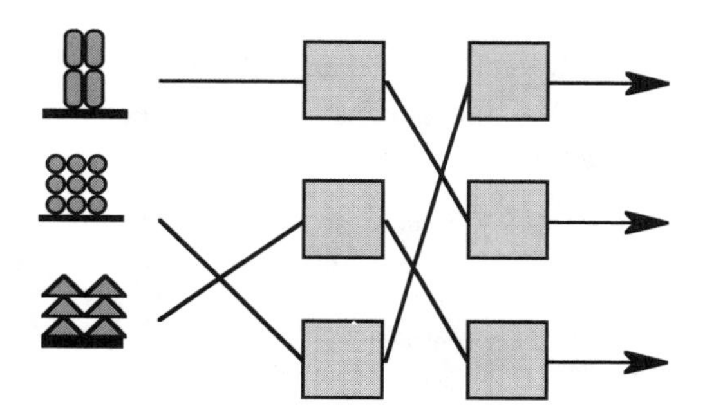

Figure 8.9 Job Shop System.

Job shops tend to be very inefficient—long lead times and high work-in-process inventories. These are some of the reasons for the inefficiencies of job shops:

- Manual material movement
- Manual operations
- Long setup times
- Low equipment utilization

During operation, the general job shop problem is to schedule the production of N jobs on M machines. For each job, the sequence of machines is known as well as the processing time on each machine. Due dates may also be known. In scheduling, four principal goals or objectives are to be achieved :

1. Minimize job lateness or tardiness
2. Minimize the flow time or time jobs spend in production
3. Maximize resource utilization
4. Minimize production costs

The decision as to which job to process next is usually based on some rule such as the following:

Rule	Definition
Shortest processing time	Select the job having the least processing time.
Earliest due date	Select the job that is due the soonest.
First come, first served	Select the job that has been waiting the longest for this workstation.
First in system, first served	Select the job that has been in the shop the longest.
Slack per remaining operation	Select the job with the smallest ratio of slack to operations remaining to be performed.

Job shops present a significant challenge due to the complexity and variability that they possess. To illustrate the magnitude of complexity of the decisions that can attend the management of a job shop, Kochan (1986) notes that:

> In a job shop with 50 machines, an average processing time of 20 min, and a backlog of 6 weeks (of) work, one might find 16,000 workpieces from which to choose the best batch to be processed. Each of these batches could have information on due date, customer, priority, batch number, number of parts, assembly identification, operation identification, physical location, and raw material requirements which might have a bearing on the choice of the best next batch to be produced.

When one adds to the batch information the processing requirements, part routings, etc., the amount of information for a given batch becomes enormous. In developing a simulation model, the modeler must be careful to reduce this information to the bare minimum necessary to make the decisions that are of importance to the flow of the jobs through the system.

Simulation traditionally has been used in job shops more for scheduling than for layout design and logistical considerations.

> Historically, simulations of job shop performance focused on the effectiveness of various queue disciplines, labor assignment rules, and dispatching criteria. Not considered in these studies specifically is how the work will be moved through a shop where machines are grouped by function, the number of personnel and other resources needed to perform the transport function, the personnel and control system to know where items are and the status of machines, and space requirements (Blackburn and Millen 1986).

Examples

- Metal working
- Fabrication or machining operations
- Maintenance facilities for aerospace industry

Performance Measures

- Time to complete a set of jobs (makespan).
- Number of completed jobs that are tardy.
- The average lateness of jobs that are completed.
- Utilization of equipment.

Decision Variables

- The job selection rule.
- The sequence in which jobs are processed.
- The routing for a particular job if alternative routings are possible.
- The resources assigned to particular jobs.
- The transfer batch size.
- Use of overlapped versus non overlapped production.
- Overtime and shift policies.
- Assignment of resources to workstations.

Questions to be Answered

- What is the optimum scheduling of jobs that minimizes makespan?
- What is the optimum scheduling of jobs that minimizes the number of tardy jobs or average tardiness?
- What is the optimum scheduling of jobs that minimizes the average flow time?
- What is the optimum batch size that minimizes the average lead time of all parts manufactured in the system?
- What is the optimum scheduling of jobs that achieves some weighted combination of the above criteria?

Model Representation

System Element	*Model Representation*
Jobs	Jobs are the entities processed in the system. Each entity represents either an individual part in a job or an entire job depending on the level of detail required. If scheduling is an objective of the simulation, each entity may be assigned an attribute to which the due date is assigned. Processing times and setup times for jobs may also be defined as job attributes.

Workstations

Workstations or machines are usually modeled as processing locations. These locations may have setup times associated with them and defined downtime and repair time distributions.

Queues

Since floorspace is usually used for queuing, there are seldom practical constraints placed on queuing. Queues are generally defined prior to each workstation and given a sufficiently large enough capacity to allow any number of jobs to be queued up in front of each workstation.

Operators

If personnel are shared, they may be modeled in the simulation as resources. If a workstation has its own dedicated operator, it may be ignored as part of the activity.

Material handling

Unless material handling is a recognized constraint, it is generally ignored. Since production is performed in batch, it can usually be assumed that when a batch finishes at one station, it is immediately available at the next station.

Routings

Routings are typically defined for each job unless there are jobs that share the same routing.

Sequencing rules

The sequencing rules are generally modeled by specifying an input selection rule for each location. The rule may be based on the value of an entity attribute representing the due date, remaining operations, etc.

Simulation Procedure

Job shops are rarely modeled from a steady-state viewpoint. The system drivers fluctuate too greatly to get any steady-state behavior. For this reason, job shop simulations usually focus on a known set of jobs to be produced. The length of the simulation is then based on the completion of all jobs rather than on a prescribed time. Multiple replications are required if random variables are included. No warm-up period should be used to obtain results. However, if there are jobs in progress at the beginning of the simulation, the model should account for them.

Since a particular situation is being simulated, often corresponding to an actual production circumstance, the beginning and ending conditions might be considered.

CELLULAR MANUFACTURING

An alternative to a job shop is cellular manufacturing in which machines are grouped into cells according to common processes. A manufacturing cell is "a group of machine tools and associated materials handling equipment that is managed by a supervisory computer" (Cutkosky, Fusser, and Milligan 1984). Manufacturing cells are often called Group Technology (GT) cells since group technology is the basis for designing the cell. *Group technology* is an approach to design based on the premise that similar things should be done similarly. Parts having similar configurations (rotational, prismatic, etc.) or similar processes (turning, grinding, etc.) should be produced by the same cell of machines. A cell is an independent group of machines but may be connected with other cells to form a flexible manufacturing system.

Manufacturing cells usually consist of from two to five CNC machine tools capable of processing a family of part types. A part family is a set of parts that require similar machinery, tooling, operations, jigs, and fixtures. The flow of parts within the cell resembles the streamlined flow achieved in line flow manufacturing. This results in greater efficiencies by consolidating groups or families of products together and treating them, from a work flow standpoint, as a single product. A cell is an excellent way to achieve the "factory within a factory" concept and is becoming a widely adopted approach to low volume, high mix manufacturing. The ideal is to have all of the operations of a job performed within a single cell. Rarely is the entire facility divided up into GT cells. The portion of the facility that remains as a functional job shop is termed the remainder cell. This usually comprises 30 to 40 percent of the total facility (Greene 1987).

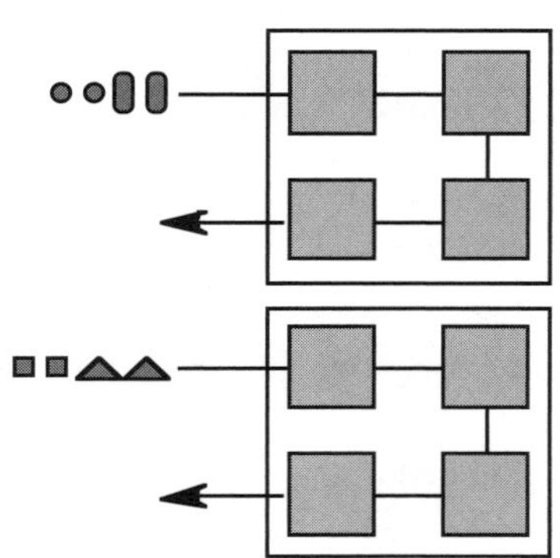

Figure 8.10 Cellular Manufacturing System.

Examples

- Serial flow cells
- Random flow cells
- Virtual cells

In a serial flow cell, all parts flow through the same sequence of machines and hence, a miniature production line is established. In random flow cells, different parts have different routings within the cell, with the effect being somewhat similar to a job shop. Machine utilization tends to be less in a random flow cell than in a serial flow cell. The concept of a virtual cell was first proposed by the National Bureau of Standards (NBS) (now the National Institute of Standards and Technology (NIST)). This concept uses a process layout of equipment just like a job shop, rather than a cellular layout. Machines are treated logically as a group even though they are physically separated. A virtual cell functions as a cell based on the needs at the time. Individual workstations are allocated to a virtual cell on a dedicated or time-sharing basis with other virtual cells. The concept of virtual cells developed from the philosophy that changing production requirements alter the part family makeup for a given production period. When the requirements alter, the allocation of individual workstations will change.

The benefits of group technology are summarized by Hollier (1980):

- Better lead times provide fast response and more reliable delivery.
- Work in process and finished stock levels are reduced.
- Output is increased because of improved resource utilization.
- Less material handling is needed.
- Better space utilization is achieved.
- Better production planning and control.
- A smaller variety of tools, jigs, and fixtures.
- Improved quality and reduced scrap.
- Simplified estimating, accounting, and work measurement.
- Improved job satisfaction, morale, and communication.
- Reduced product design variety.

The main disadvantages are:

- Implementation costs.
- Rate of change in product range and mix.
- Difficulties with out-of-cell operations.
- Coexistence with non-cellular systems.

Performance Measures

- Machine utilization

- Production rate
- Utilization of the operator or robot
- Utilization of the bottleneck station

Decision Variables

- The number and types of machines in the work cells
- The batch size of a particular part type
- Sequencing of part types within the cell
- Material handling priorities within the cell

Questions to be Answered

- How many of each type of machine are required to balance production?
- What is the best cell design for maximizing throughput?
- What is the utilization of the bottleneck machine given a particular sequence of orders?
- What is the optimum sequencing of part types through the cell that minimizes setup?

Model Representation

System Element	*Model Representation*
Parts	The parts are the entities in the model. If steady-state behavior is being analyzed, the entire part family may be represented by a single entity type whose routing and processing times are probabilistically determined based on the distribution of part types within the family.
Workstations	Workstations or machines are typically represented by the route locations in the model. These locations may have setup times associated with them and defined failure and repair time distributions.
Queues	There are seldom queues in a manufacturing cell since the number of machines in a cell is small and the goal is to keep a steady flow of product through the cell.
Operators	Since machines in a cell are frequently unmanned, there are often no operators to model. If an operator is used, it is usually to load and unload the machines in the cell, in which case the operator functions as a material handler described below.

Material handling

If a robot is used for handling, it can usually be modeled as a general use resource although a robot construct may be used if available. For cell performance analysis, it is not necessary to model the kinematics of the robot as long as the time to move material from one station to another is known. Additionally, reliability and maintenance factors may need to be considered for steady-state simulations.

If material handling is manual, it is easily modeled using a general resource construct. Since humans are not as consistent as mechanisms, there needs to be variability factored into the handling times.

Regardless of the handling method used, it is usually helpful to model the task priority rules for the handler to test whether the best operating policies are being used.

Routing

Part routings within a cell are generally quite straightforward. There may be alternative machine selections which are easily modeled using location selection rules.

Sequencing rules

The sequencing rules are generally modeled by specifying either an input selection rule for each location, or by specifying a sequencing operation at the beginning that all parts pass through to determine the order of production. In cellular production, the sequencing rule is frequently based on minimizing machine setup and maximizing resource utilization rather than considering due date factors. For this reason, usually more detailed if-then logic is needed to make the decision.

Simulation Procedure

Manufacturing cells are often modeled with the same objectives as a job shop simulation: to determine the best sequencing of jobs and to maximize resource utilization. In this case, a known set of jobs is modeled and no steady-state behavior is analyzed. It is sometimes desirable to predict potential bottlenecks that may exist on an ongoing basis by modeling an estimated mix of jobs to be produced.

Rather than model a predetermined set of jobs, probabilities are used that correspond to the likelihood that the next job will be of a particular type.

FLEXIBLE MANUFACTURING SYSTEMS

A flexible manufacturing system (FMS) is a network of NC machines, machining centers, and/or cells with automated storage and transportation systems for part and tool storage and transport. A flexible manufacturing system is "a computer controlled configuration of semi-independent workstations and a material handling system designed to efficiently manufacture more than one kind of part at low to medium volumes" (*FMS Handbook* 1983). The workstations in an FMS usually consist of numerically controlled machines with input and output buffers for limited queuing. Central storage locations are sometimes used such as an automated storage/retrieval system (AS/RS) or a carousel (see Figure 8.11).

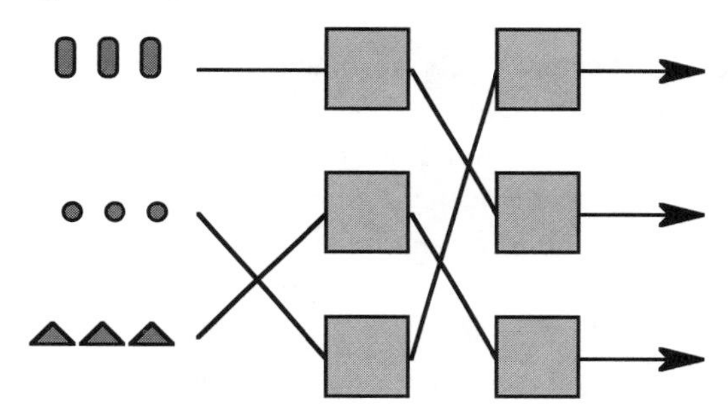

Figure 8.11 Flexible Manufacturing System.

An FMS is typically a system in which parts are allowed random access to any workstation in the system through the use of a flexible material handling system. Typically, an FMS utilizes one of the following three types of material handling systems:

- Conveyors
- Automated vehicles
- Robots

Conveyors are generally recirculating with a spur or loop at each station. Conveyors are not used as frequently now as in the past, largely due to the expense and difficulty in reconfiguring them once installed. One type of automated vehicle used in an FMS is a cart-on-track in which a cart traverses back and forth on a linear track. This type of system is used for relatively small FMSs. A more popular and more versatile vehicle is an AGV (automatic guided vehicle) which follows inductive signals emitted from electric wires (called a guide path) imbedded in the floor. Some AGVs have inertial guidance systems that make them more versatile and easier to install.

Robots used in FMS's consist of mobile robots travelling on rails on the floor or on overhead gantries. Robots are sometimes mounted on AGV's to provide both delivery and loading/unloading, especially of tools into and out of tool magazines on CNC machines.

Since the material handling system is usually automated, parts are loaded onto pallets at a load station to facilitate handling. *Fixtures* are used to position parts on the pallet which is usually designed for a particular part or family of similar parts. Fixtures may either be permanently bolted to the pallet or removable so that other parts may use the pallet. If the parts are small, several parts of the same or different type may be grouped together on a single pallet. There may be one or more parts to a fixture and one or more fixtures to a pallet. Fixturing is usually a manually performed operation that utilizes a mechanical assist device such as a hoist if parts are heavy. After being fixtured, the palletized parts move via a material handling system to queues at the workstations. When the workstation (usually a machine tool) becomes available, palletized parts transfer from the queue to the workstation (sometimes via a shuttle system) and then, after the operation is performed, they transfer to an output queue to await delivery to the next operation (see *FMS Handbook* 1983). Parts may also require more than one fixturing or "setup" during processing.

While many of the operations in an FMS are automated, most FMS's still require the use of operators for loading/unloading parts, tool preparation, performing planned and unplanned maintenance, etc.

An FMS offers tremendous benefits such as higher machine utilization, lower unit costs, shorter lead times, higher quality, and quicker response to market changes. These benefits are only realized, however, when careful and effective planning accompanies the design and implementation of an FMS. An FMS is a very complex system with interconnected components and interrelated operations all controlled and integrated under a sophisticated control system. If not carefully designed with consideration given to the integration requirements of the system, failure is inevitable.

The main purposes of simulation of FMSs are (Carrie 1988):

- Assess the capacity and equipment utilizations in the system.
- Identify the bottlenecks.
- Compare the performance of alternative designs.
- Ensure that there are no fundamental weaknesses in the FMS design.
- Develop operating strategies for work scheduling and job sequencing.

A detailed listing of data required for modeling an FMS can be found in Bevans (1982). So vital is simulation to the design of FMS that one European supplier of FMS simply states: "Without it , there is no FMS."

Performance Measures

- Material handling system utilization
- Machine utilization
- Production rate
- Makespan

Decision Variables

- The number and types of machines in the system.
- The production batch size.
- The move batch size.
- Sequencing of part types through the system.
- Material handling priorities.
- Number and type of fixtures available.
- Part types capable of using each fixture type.
- Tool requirements for each operation.
- Buffer size.

Questions to be Answered

- How many of each type of machine are required to meet production requirements?
- Where are the bottlenecks in the system?
- What is the optimum sequencing of products through the system?
- What should buffer capacities be to minimize idle time?
- How many material handling units (robots or AGVs) are needed to meet production requirements

Model Representation

System Element	***Model Representation***
Workpieces fixtures, pallets, tooling	The workpieces, fixtures, pallets and tooling may be all modeled using entities. Workpieces are the parts being produced. Fixtures are used to position parts on pallets. They are usually designed for a particular part or family of similar parts. They may also be permanently bolted to the pallet or removable so that other parts may use the pallet. If the parts are small, several parts of the same or different type may be grouped together on a single pallet.

Fixtures, pallets and tooling are confined to the system. If steady-state behavior is being analyzed, the entire part family may be represented by a single entity type whose routing and processing times are probabilistically determined based on the distribution of part types within the family.

Workstations — Workstations or machines are usually modeled as processing locations. These locations may have setup times associated with them and defined failure and repair time distributions.

Queues — The primary queues are the input and output queues at each station. These are modeled using a buffer or queuing construct.

Operators — The operator may be modeled as general-use resources that may be assigned to one or more workstation.

Routing — Part routings within an FMS can be quite complex. There are often alternative station selections which require the use of special selection rules.

Control system — Since an FMS is computer controlled, the control logic to be modeled can be quite complex. Often detailed, "if-then" logic may need to be used.

Inspection — Inspection locations and frequencies must be included as part of processing logic.

An FMS has material handling requirements within each cell in the system that are similar to manufacturing cells. Movement between cells becomes a complex part of the model due to the complexity of the configuration and control logic. If possible, it is preferable to utilize AGV or conveyor constructs to model this complexity. Task prioritizing and other complex decision logic will need to be modeled.

Simulation Procedure

If the system is cleared of parts periodically such as at the end of each shift, day or week, it may be possible to get a steady-state behavior. Otherwise a transient analysis should be performed.

BATCH FLOW SHOP

A batch flow shop, or flow shop, utilizes a product layout in which a sequence of workstations is visited in the same sequence by different product batches. Batch flow shops are similar to job shops in that many different types of discrete parts are produced in batches. In a batch flow shop, however, all flow is basically unidirectional following the same route through the manufacturing facility in a production line fashion.

A batch flow shop is composed of one or more production lines which support a batch flow of parts. Each part may not require processing at each station and may even bypass some stations, but the same general flow is followed. All operations on parts of the same type are usually performed in the same order. Batch flow shops are most commonly found in the textile industry in which batches of different styles and sizes are processed through the same sequence of operations. In flow manufacturing, the emphasis tends to be on efficiency and streamlining flow. This means getting rid of bottlenecks and making effective use of resources (see Figure 8.12).

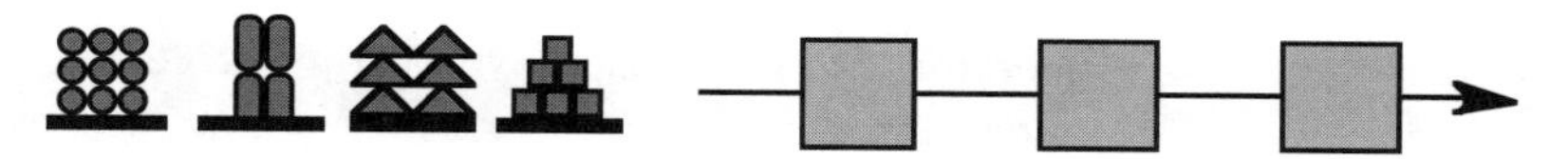

Figure 8.12 Batch Flow Shop.

Performance Measures

- Resource utilization
- Throughput capacity
- Work in process
- Cost

Decision Variables

- Queuing between stations
- The production batch size of a particular part type
- The move batch size
- The sequence of products

Questions to be Answered

- How many of each type of machine are required to balance production?
- What is the bottleneck machine?
- Should overlapped production be used?
- What is the best sequence of products that minimizes changeover?
- Should changeover be performed as stations are finished or after the entire line is emptied?

Model Representation

System Element	*Model Representation*
Parts	The parts are the entities in the model. If overlapped batching is not used, the entire move batch may be modeled as a single entity. In this case, the operation time may be the sum of all of the individual part operation times.
Workstations	Workstations or machines are typically represented by the route locations in the model. These locations frequently have setup times associated with them and defined failure and repair time distributions.
Buffers	Buffers are modeled as queues and are important to be modeled since they directly affect resource utilization and balance of flow.
Operators	Operators are often used even if only to load and unload the machines or perform machine setups. Usually, however, the operator is dedicated to a specific station so that it may not be necessary to include in the model.
Operations	Receiving, cutting, batching, batch operations (heat treat, plating), machining, assembly, test, rework, packing, and shipping.
Material handling	In a batch flow shop, the material handling system can often be ignored since the flow is from one location to the next with adequate queuing to prevent any machine idle time due to inefficient handling.
Routing	Part routings within a batch flow shop are easy to define in terms of the routing sequence. More difficult is to accurately define the input/output quantity relationships. It is important that the quantities reflect the consolidation made for movement and the individualized processing.

Simulation Procedure

Batch flow shops are usually modeled to obtain steady-state behavior since a repeating sequence of products is always being produced. Because production occurs in a defined cycle, it may be useful to analyze the results over the cycle instead of, or in addition to, the overall system behavior. Simulation is usually based on time although it could be based on the ending of a production cycle.

LINE FLOW MANUFACTURING (PRODUCTION/ASSEMBLY LINES)

Line flow manufacturing consists of production/assembly lines and transfer lines in which products move and are processed individually rather than in batch. Line flow systems are characteristic of many mass production operations in which workstations are set up by product in a serial arrangement and dedicated to manufacturing or assembling a single product. The idea is to achieve a streamlined, continuous flow of material that leads to maximum productivity. Labor and machines are highly utilized with little idle time. Since transfer lines are sufficiently different from production/assembly lines, they are treated separately.

Production and assembly operations that are of a line flow type are comprised of a serial combination of two or more production, assembly, and packaging stations typically connected by a continuous material handling system such as a conveyor. Non-synchronous conveyors have become the most popular and efficient material handling system because they permit parts to maintain a continuous flow while still allowing them to queue up when necessary. Operations are often performed by hand and therefore present a special challenge to keep the flow as continuous as possible. This is achieved by balancing the work load among stations, keeping each station busy, and reducing the variability of each operation. Line balancing heuristics for determining the number and work content of workstations are relatively easy to implement. Unfortunately, they do not take into account the variability that frequently exists in operation times. Simulation accounts for operation variability and is a more accurate predictor of production capacity.

Of course, changing the number and work content of workstations is only possible to the extent that task elements permit. The more stations, the lower the cycle time, hence the higher the throughput. An alternative to stretching out a line into more stations to increase throughput is to add parallel lines. At one extreme, a single line consisting of n serial workstations may be used. The job is broken down into as many small subtasks as possible without overproducing. At the other extreme would be n parallel lines consisting of a single station each. The entire job is performed on each single station line with as many lines as are needed to meet demand. Many lines lie somewhere in between these two extremes and consist of a mix of serial and parallel stations.

The placement and size of buffers has an impact on inventory costs and system throughput. If the entire line stopped every time a part was unavailable or a station failed, the line would be going down all the time. Buffers allow workstations to operate independently thus cushioning the effects of scrap, part shortages, unequal production rates, workstation failures, or operator delays.

Many products are not produced in sufficient quantities to justify a dedicated line. Frequently a production or assembly line is used to produce a family of similar products. Products are produced in batch runs in which the line is temporarily shut down for product changeovers while machine adjustments are made for the next product. Recent attention to just-in-time techniques has led to the use of *mixed lines,* in which flexible automation and quick-setup procedures allow smaller batch sizes to be run. As long as the input mixture matches product demand rates, the precise sequence of products introduced into the line should attempt to minimize workstation imbalances between model types.

Production and assembly lines may be either *paced* in which movement occurs at a fixed rate and the operator must keep pace with the line, or they may be *unpaced* in which the rate of flow is determined only by the speed of the worker (see Figure 8.13).

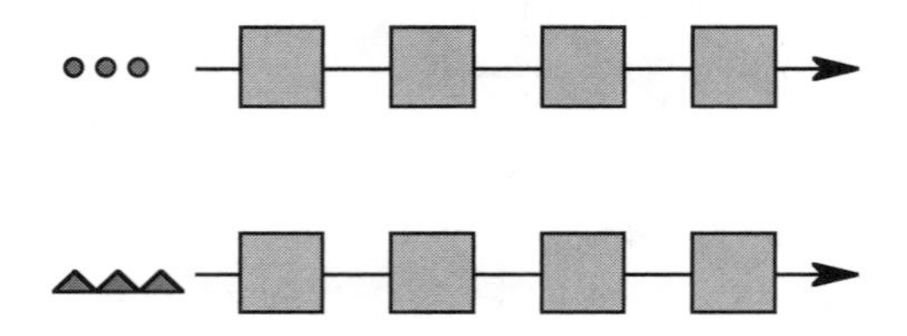

Figure 8.13 Production/Assembly Line System.

Examples

- Appliances
- Consumer products
- Medical instruments

Performance Measures

- Average and variation in throughput capacity.
- Average and variation in work in process.
- Cost.
- Balance delay (sum of the idle time for all station/sum of the scheduled time for all stations).
- System efficiency (actual throughput of the system/theoretical throughput capacity of the slowest station).
- Percentage of time stations are blocked.
- Percentage of time stations are starved.

Decision Variables

- The number and types of machines in the work cells
- The production batch size of a particular part type
- Type of material handling
- The sequence of products
- Number of stations
- Placement and size of buffers

Questions to be Answered

- How many stations are required to meet production?
- What is the bottleneck machine?
- What is the best way to minimize changeover time?
- How does workstation reliability affect system performance?
- What impact does interstation buffering have on system performance?
- How much buffering is needed to achieve a particular production rate?

Model Representation

System Element	***Model Representation***
Parts	The parts are represented by entities.
Workstations	The workstations are the route locations in the model. If a series of workstations collectively produce at a constant rate, they may be represented as a single location. Production line locations frequently have setup times associated with them and defined failure and repair time distributions.
Buffers	Buffers exist only in paced lines. They are modeled by using a queue or multi-capacity location.
Entity processes	For production lines, processes are easily defined in terms of the operation time. For assembly lines, there may be a need to model the component part availability. This is done using a construct that permits the joining of a component part to a main part.
Operators	Operators are often used even if just to load and unload the machines or to perform machine setups. Usually, the operator is dedicated to a specific station—it may not be necessary to include the operator in the model.

Material handling	In a line flow system, the material handling system can be a critical factor, especially if it is a conveyor which constrains the quantity of parts that can be moved between any two locations.
Routing	Part routings within a line flow system are usually very straightforward, with parts moving serially through the locations. Repair loops may complicate the routing, requiring the use of probabilistic routings.

Simulation Procedure

In simulating a line flow system, we are generally interested in the steady-state behavior of the system. Consequently, we must determine a suitable warm-up period followed by a lengthy run time that produces a sample that is representative of the steady-state. If precision is important, some means of obtaining multiple samples (multiple replications, batch means, etc.) needs to be used.

If the system is a mixed model line, there may be several steady-state conditions depending on the current product being produced.

LINE FLOW MANUFACTURING (TRANSFER LINES)

Transfer lines are used predominantly in automotive and heavy equipment manufacturing. A transfer line is very similar to an assembly line except the operations are usually machining operations and the stations are automated. It is essentially a paced, serial machining line in which stations are subject to various types of breakdowns. The goal is essentially the same as that for an assembly line: keep stations working through a combination of ensuring that equipment and tooling are reliable and repair time is kept to a minimum. Tool change policies in a transfer line can have a significant impact on system performance.

Transfer lines may be synchronous or nonsynchronous (also referred to as asynchronous). A *synchronous line* is one in which all parts transfer simultaneously when the station with the longest cycle time has completed its cycle. A *nonsynchronous line* provides banking or buffering between stations to decouple the workstations so the impact of cycle time variations and station interruptions are minimized. In a nonsynchronous line, when a station completes its operation, it passes the entity forward into the next buffer. If the buffer is full, the feeding station becomes blocked and is unable to produce any more parts. Sometimes when buffers fill up, parts are removed by hand to an offline holding area to keep stations from becoming blocked. Some transfer lines have both synchronous and nonsynchronous sections (see Figure 8.14).

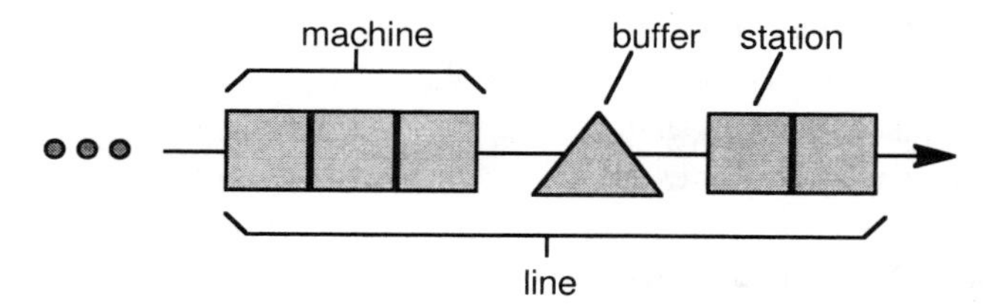

Figure 8.14 Transfer Line System.

Originally, a transfer line was considered to be an example of "hard" or "fixed" automation. Even minor changes in product design required major equipment modifications. A new product rendered the entire line obsolete. With the development of programmable controllers in recent years, transfer lines are becoming flexible and are considered a type of flexible manufacturing system.

One issue in material handling deals with finding the optimum number of pallets in a closed, nonsynchronous pallet system. Such a system is characterized by a fixed number of pallets that continually recirculate through the system. These pallets are usually loaded at one station and unloaded at another station located just prior to the load station.

Obviously, the system should have enough pallets to fill every workstation in the system. On the other hand, if every nonworkstation position is filled, the system will lock up. Generally, productivity increases as the number of pallets increases up to a certain quantity, beyond which productivity begins to decrease. The optimal quantity tends to be close to the sum of all of the workstation positions plus one-half of the buffer pallet positions.

A typical goal of transfer lines is to find buffer sizes that will assure that the line is available or productive 90 to 95 percent of the time. Sometimes the goal is to provide a sufficient buffer to protect the operation against the longest tool change time of a downstream operation.

Examples

- Engine block machining line
- Transmission case machining line

Performance measures

- Resource utilization
- Throughput capacity
- Work in process
- System efficiency (actual throughput of the system / throughput capacity of the slowest station at 100 percent efficiency)

Decision Variables

- The number and types of stations in the line
- Buffer placement and sizing
- Maintenance procedures

Design and Operational Issues

- What is the bottleneck machine?
- What advantage, if any, does preventive maintenance and periodic tool replacement have over corrective maintenance and tool replacement upon failure?
- What is the optimum number of maintenance personnel to have on a line?
- In a closed-loop, nonsynchronous, palletized transfer line, what is the optimum number of pallets?

Model Representation

System Element	***Model Representation***
Parts, fixtures pallets	The parts, fixtures and pallets are the model entities.
Stations	Stations may be modeled individually or collectively, depending on the level of detail required in the model. Often, a series of stations can be modeled as a single location.
Buffers	Buffers are modeled by using queues or multi-capacity locations.
Entity processes	Operations in a transfer machine can be modeled as a simple operation time if an entire machine or block of synchronous stations is modeled as a single location. Otherwise, the operation time specification is a bit tricky since it is dependent on all stations finishing their operation at the same time. One might initially assign the time of the slowest operation to every station. Unfortunately, this does not account for the synchronization of operations. Usually, sychronization requires a timer to be set up for each station which represents the operation for all stations.
Operators	It is often unnecessary to model operators unless they constrain the operation of the line in some way.

Material handling	In a transfer line, the material handling system is integrated with the machine in the form of a transfer mechanism such as a lift-and-carry device.
Breakdowns	Breakdowns are usually defined by station based on number of entities processed at the station. The number of cycles between failures often follows a geometric distribution. Breakdowns are of two types: individual station breakdowns and entire machine breakdowns.
Routing	Part routings are easy to define since parts move sequentially through a series of stations.

Simulation Procedure

Transfer line simulation follows the same procedure used for simulating production and assembly lines as described earlier in this chapter.

CONTINUOUS PROCESS SYSTEMS

A continuous process system is not a type of discrete part manufacturing system, but it has characteristics that are similar enough to warrant mention here. Continuous processing involves the production of bulk substances or materials such as chemicals, liquids, plastics, metals, textiles and paper (see Figure 8.15).

Figure 8.15 Continuous Process System.

Pritsker (1986) presents a simulation model in which the level of oil in a storage tank varies depending on whether oil is being pumped into the tank or oil is being pumped out of the tank. If the rate of flow into or out of the tank is constant, then the change in level is the rate of flow multiplied by the length of time. If, however, the rate of flow varies continuously, it is necessary to integrate the function defining the rate of flow over the time interval concerned (see Figure 8.16).

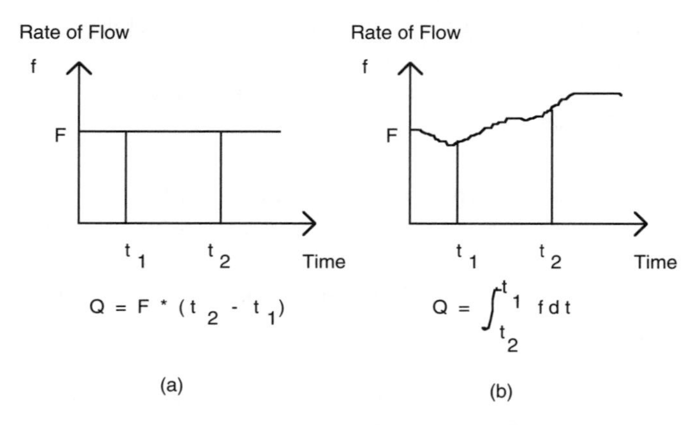

Figure 8.16 Quantity of Flow When (a) Rate is Constant, and (b) Rate is Changing.

In continuous modeling, a time step needs to be defined for advancing the clock. With a smaller time step, more precision is obtained, but the model runs slower because of the frequency of computations. The larger the time increment, the less precise the outcome, but the model will run quicker.

Examples

- Mills (paper, textile, etc.)
- Chemical processing
- Refineries

SUMMARY

This chapter focused on the issues and procedures for modeling manufacturing systems. Different types of manufacturing systems were discussed with examples given of each. Suggestions were offered regarding the way in which certain system elements might be represented in a model.

Chapter 9
Modeling Material Handling Systems

"Material handling uses the right method to provide the right amount of the right material, at the right place, at the right time, in the right sequence, at the right cost."

James A. Tompkins

INTRODUCTION

Material handling systems utilize resources to move entities from one location to another. While material handling systems are not uncommon in service systems, their main use is in manufacturing systems. Material handling frequently accounts for the majority of the production activity. On the average, 50 percent of companies' operation costs are composed of material handling costs (Meyers 1993). Given the impact of material handling on productivity and operation costs, it is crucial that material handling decisions are based on sound principles and practices.

This chapter examines simulation techniques for modeling material handling systems. Material handling systems represent one of the most complicated, yet in many instances, the most important element in the model. Conveyor systems and automatic guided vehicle systems often provide the backbone for the activity that takes place. Since a basic knowledge of material handling technologies and decision variables is essential to modeling material handling systems, we will briefly describe the operating characteristics of each type of material handling system.

MATERIAL HANDLING PRINCIPLES

It is often felt that material handling is a necessary evil and that the ideal system is one in which no handling is needed. In reality, material handling can be a tremendous support to the production process if performed intelligently.

Following is a list of twenty principles developed by the College-Industry Council on Material Handling Education (CICMHE) as a guide to designing or modifying material handling systems:

1. *Planning Principle.* Plan all material handling and storage activities to obtain maximum overall operating efficiency.
2. *System Principle.* Integrate as many handling activities as is practical into a coordinated system of operations: vendor, receiving, storage, production, inspection, packaging, warehousing, shipping, transportation, and customer.
3. *Material Flow Principle.* Provide an operation sequence and equipment layout optimizing material flow.
4. *Simplification Principle.* Simplify handling by reducing, eliminating, or combining unnecessary movements and/or equipment.
5. *Gravity Principle.* Utilize gravity to move material wherever practical.
6. *Space Utilization Principle.* Make optimum utilization of building cube.
7. *Unit Size Principle.* Increase the quantity, size, or weight of unit loads or flow rate.
8. *Mechanization Principle.* Mechanize handling operations.
9. *Automation Principle.* Provide automation to include production, handling, and storage functions.
10. *Equipment Selection Principle.* In selecting handling equipment, consider all aspects of the material being handled—the movement and the method to be used.
11. *Standardization Principle.* Standardize handling methods as well as types and sizes of handling equipment.
12. *Adaptability Principle.* Use methods and equipment that can best perform a variety of tasks and applications where special purpose equipment is not justified.
13. *Dead Weight Principle.* Reduce ratio of dead weight of mobile handling equipment to load carried.
14. *Utilization Principle.* Plan for optimum utilization of handling equipment and manpower.
15. *Maintenance Principle.* Plan for preventative maintenance and scheduled repairs of all handling equipment.
16. *Obsolescence Principle.* Replace obsolete handling methods and equipment when more efficient methods or equipment will improve operations.
17. *Control Principle.* Use material handling activities to improve control of production inventory and order handling.

18. *Capacity Principle.* Use handling equipment to help achieve desired production capacity.
19. *Performance Principle.* Determine effectiveness of handling performance in terms of expense per unit handled.
20. *Safety Principle.* Provide suitable methods and equipment for safe handling.

MATERIAL HANDLING CLASSIFICATION

With over five hundred different types of material handling equipment to choose from, the selection process can be quite staggering. We shall not attempt here to describe or even list all of the possible types of equipment. Since our primary purpose is modeling of material handling systems, we are only interested in the general categories of equipment with common operating characteristics. Material handling systems are traditionally classified into one of the following categories (Tompkins and White 1984):

- Conveyors
- Industrial Vehicles
- Automated Storage/Retrieval Systems
- Automatic Guided Vehicle Systems
- Cranes and Hoists
- Robots

Missing from this list is "hand carrying" which is still widely practiced if for no other purpose than to load and unload machines. From a simple modeling standpoint, we could classify all material handling as being either conveyors that provide continuous movement along a fixed path, or transporters that provide intermittent movement of loads. For more detailed modeling, or to eliminate the need to convert motion along multiple axes to a single move time, it may be desirable to have access to modeling constructs that are designed for the specific type of system being modeled.

Classifying a particular handling device as being either a transporter or a conveyor is not always straightforward. The distinction is obvious when the device is generally a fixed device along which parts flow, such as a belt conveyor, or when the device is mobile which intermittently moves parts, such as a fork truck. The distinction becomes difficult for some material handling systems, however, such as monorail systems which consist of multiple mobile devices running along a fixed track. Such a system produces the effect of continuous part flow even though parts are actually being moved individually by each device. From a modeling standpoint, it does not really matter how they are classified as long as the movement produced is accurately represented by the model. From purely a modeling efficiency standpoint, it would be easier to model a semicontinuous flow system using the continuous flow algorithm of a conveyor than to try to break it down into a large number of independent transport devices. Since the entity being moved is often based on the unit load, we will use the term *load* to define the entity of movement.

CONVEYORS

A conveyor is a track, rail, chain, belt, etc. that provides continuous movement of loads over a fixed path. Conveyors are generally used for high volume movement over short to medium distances. Some overhead or towline systems move material over longer distances. Overhead systems most often move parts individually on carriers, especially if the parts are large. Floor-mounted conveyors usually move multiple items at a time in a box or container. Conveyor speeds generally range from 20 to 80 feet per minute (fpm) with high speed sortation conveyors reaching speeds of up to 500 fpm in general merchandising operations.

Conveyor Types

Conveyors are sometimes classified according to function or nature of operation such as the following:

- *Transport conveyor.* Transports loads from one point to another.
- *Sortation conveyor.* Provides diverting capability to divert loads off the conveyor.
- *Recirculating conveyor.* A conveyor loop that causes loads to recirculate until they are ready for diverting.
- *Branching conveyor.* A branching conveyor is characterized by a main line with branches or spurs on which parts merge on to the main line conveyor or on which parts divert off of the main line. Branches may, in turn, have additional spurs or branches.

There are many different types of conveyors based on physical characteristics. Among the most common types are the following:

- Belt
- Roller
- Chain
- Trolley
- Power and free
- Towline
- Monorails

Belt conveyors. Belt conveyors have a circulating belt on which loads are carried. Since loads move on a common span of fabric, if the conveyor stops, all loads must stop.

Roller conveyors. Roller conveyors utilize rollers that may be "dead" or "live". Dead rollers are placed on a grade to use the effect of gravity on the load to actuate the rollers. Many roller conveyors utilize "live" or powered rollers to convey loads. Roller conveyors are often used for simple load accumulation.

Chain conveyors. Chain conveyors are much like a belt conveyor in that one or more circulating chains convey loads. Chain conveyors are sometimes used in combination with roller conveyors to provide right angle transfers for heavy loads. (A pop-up section allows the chains running in one direction to "comb" through the rollers running in a perpendicular direction.)

Trolley conveyors. A trolley conveyor consists of a series of trolleys (small wheels) that ride along a track and are connected by a chain, cable or rod. Load carriers are typically suspended from the trolleys. The trolleys, and the load carriers, continually move: the entire system must be stopped to bring a single load to a stop. No accumulation can take place on the conveyor itself. Parts are frequently loaded and unloaded onto trolley conveyors while the carriers are in motion (usually not faster than 30 fpm). Because a trolley conveyor can be designed with a vertical dip, it can descend to automatically capture a carrier. Automatic unloading is simpler. A carrier can be tipped to discharge a load or a cylinder can be actuated to push a load off a carrier onto an accumulation device. Discharge points can be programmed.

Power-and-free conveyors. Power-and-free conveyors consist of two parallel trolley systems, one of which is powered while trolleys in the other system are nonpowered and free moving. Loads are attached to the free trolleys and a "dog" or engaging pin on the powered trolleys engages and disengages the passive free trolleys to cause movement and stoppage. In some power and free systems, engaged load trolleys will automatically disengage themselves from the power line when they contact carriers ahead of them. This allows loads to be accumulated. In the free state, loads can be moved by hand or gravity; they can be switched, or banked, or merely delayed for processing.

Tow conveyors. A tow conveyor or towline system is an endless chain running overhead in a track, or on or beneath the ground in a track. The chain has a dog or other device for pulling (towing) wheeled carts on the floor. Towlines are best suited for applications where precise handling is unimportant. Tow carts are relatively inexpensive compared to powered vehicles; many of them can be added to a system to increase throughput and be used for accumulation.

An underfloor towline uses a tow chain in a trough under the floor. The chain moves continuously and cart movement is controlled by extending a drive pin from the cart down into the chain. At specific points along the guide way, computer-operated stop mechanisms raise the drive pins to halt cart movement. One advantage of this system is that it can provide some automatic buffering with stationary carts along the track.

Towline systems operate like power-and-free systems and, in fact, some towline systems are simply power-and-free systems that have been inverted. By using the floor to support the weight, heavier loads can be transported.

Monorail conveyors. Automatic monorail systems have self-powered carriers that can move at speeds of up to 300 fpm. In this respect, a monorail system is more of a discrete unit movement system than a conveyor system. Travel can be disengaged automatically in accumulation as the leading "beaver tail" contacts the limit switch on the carrier in front of it (Schwind 1988).

Conveyor Load Control

Load control systems monitor and direct the movement of loads on the conveyor. Conveyors range from little, if any, load control to complete computerized control. For simple accumulation conveyors, for example, no control is needed since all loads do exactly the same thing—they move to the end where they form a queue. In conveyor systems where there are single or multiple pickup points and multiple drop-off points, some type of load control system is necessary. Load control is usually based on either some type of an escort memory device or on realtime computer tracking.

Escort memory. Escort memory is based on some type of an identifier that is either affixed to or accompanies the load. This identifier might be a barcode, magnetic strip, etc. that determines the destination of the load. Some conveyor systems utilize programmable escort memory devices that control where loads branch and stop. For example, a power-and-free carrier might have a signal pin or magnetic probe scheme which can be set remotely and read remotely. Towline carts commonly use a magnetic probe that is manually transferred to change codes, a plastic probe that incorporates a reflector read by electric eyes mounted below the floor, or by a barcode holder with a removable barcode read by an optical scanner. Automatic overhead conveyors can have a moveable signal pin arrangement detectable by limit switches or sometimes by proximity switches.

Escort memory systems are simple and inexpensive and are generally adequate for most material handling requirements.

Computer control. Computer control maintains all of the information about the load, or at least the destination of the load, in computer memory. In continuous load tracking, the computer knows the load ID and tracks by zone every move of the load to see that it gets to the right destination. In intermittent tracking, the computer knows the destination for each load ID and needs only determine the load ID to direct the movement of the load. In this situation, the computer checks the load ID at each decision point and, knowing the destination of the load, sends a control signal directing the load.

Computer control requires more careful operation and management than an escort memory system, especially if it loses track of a load. One advantage however, is that computerized tracking allows dynamic reassignment of a load destination much more easily than an escort memory system.

Operational Characteristics

The operational characteristics of conveyor systems are entirely unlike those of other material handling systems. Some of the main distinguishing features pertain to the nature of load transport, how capacity is determined, and how entity pickup and delivery occurs.

Load transport. A conveyor spans the distance of travel along which loads move continuously while other systems consist of mobile devices that move one or more loads as a batch move.

Capacity. The capacity of a conveyor is determined by the speed and load spacing rather than having a stated capacity. Specifically, it is a function of the minimum allowable interload spacing on (which is the length of a queue position in the case of accumulation conveyors) and the length of the conveyor. In practice this may never be reached because the conveyor speed may be so fast that loads are unable to transfer onto the conveyor at the minimum allowable spacing. Furthermore, there may be an intentional control imposed on the number of loads that are permitted on a particular conveyor section.

Entity pickup and delivery. Conveyors usually do not pick up and drop off loads as in the case of lift trucks, pallet jacks, or hand carrying; loads must be placed onto and removed from conveyors. Putting a load onto an input spur and identifying the load to the system so it can enter the conveyor system is commonly referred to as load induction (the spur is called an induction conveyor).

Modeling Conveyor Systems

Depending on the nature of the conveyor and its operation, modeling a conveyor can be either quite straightforward or extremely complex. For simple single conveyor sections, modeling is very simple. Conveyor networks, on the other hand, give rise to several complexities (recirculating, merging, etc.) that make it nearly impossible to predict how the system will perform. These types of conveyors make especially good candidates for simulation.

From a modeling standpoint, a conveyor may be classified as being either accumulation or non-accumulation, with either fixed or random load spacing (see Table 9.1).

Table 9.1 Types of Conveyors for Modeling Purposes

	Accumulation	Non-accumulation
Fixed Spacing	Power and Free	Trolley
	Towline	Sortation
Random Spacing	Roller	Belt
	Monorail	Chain

Accumulation conveyors. Accumulation conveyors are a class of conveyors that provide queuing capability—the lead load stops, succeeding loads still advance along the conveyor and form a queue behind the stopped part. Accumulation conveyors permit independent or non-synchronous part movement and therefore maximize part carrying capacity.

Non-accumulation conveyors. Non-accumulation or simple transport conveyors are usually powered conveyors having a single drive—all movement on the conveyor occurs in unison or synchronously. If a load must be brought to a stop, the entire conveyor and every load on it is also brought to a stop. When the conveyor starts again, all load movement resumes.

Fixed spacing. Conveyors with fixed spacing, also referred to as segmented conveyors, require that loads be introduced to the conveyor at fixed intervals. Tow conveyors and power-and-free conveyors which have fixed-spaced dogs that engage and disengage a part-carrying cart or trolley are examples of accumulation conveyors with fixed-load spacing. Trolley conveyors with hanging carriers at fixed intervals or tray conveyors with permanently attached trays at fixed intervals are examples of non-accumulation conveyors with fixed spacing.

Random load spacing. Random load spacing permits parts to be placed at any distance from another load on the same conveyor. A powered roller queuing conveyor is a common example of an accumulation conveyor with random load spacing. Powered conveyors typically provide non-contact accumulation while gravity conveyors provide contact accumulation (the loads accumulate against each other). A powered belt conveyor is a typical example of a non accumulation conveyor with random spacing. Parts may be interjected anywhere on the belt and when the belt stops, all parts stop.

Performance measures. Performance measures in conveyor system simulation are as follows:

- Load rate capacity
- Delivery time
- Queue sizes (for accumulation conveyors)

Decision variables. Several issues are addressed in conveyor analysis:

- Conveyor speed
- Accumulation size
- Number of carriers

Questions to be answered. The following are common questions that simulation can help answer in designing and operating a conveyor system:

- What is the minimum conveyor speed that still meets rate requirements?
- What is the load rate capacity of the conveyor?
- What is the load delivery time for different activity levels?
- How much queuing is needed on accumulation conveyors?
- How many carriers are needed on a trolley or power-and-free conveyor?
- What is the optimum number of pallets that maximizes productivity in a recirculating conveyor?

Modeling Single Section Conveyors

Conveyors used for simple queuing or buffering often consist of a single stretch or section of conveyor. Loads enter at one end and are removed at the other end. Load spacing is generally random. These types of conveyors are generally quite easy to model, especially using simulation products that provide conveyor constructs. The modeler merely defines the length and speed and states whether the conveyor is an accumulation conveyor or non-accumulation conveyor.

If the simulation product being used does not have a conveyor construct, or if there is so much activity on one or more conveyor sections that the model begins to run sluggishly, it may be necessary to model the conveyor using a more primitive approach that is efficient yet maintains reasonable accuracy. Henriksen and Schriber (1986) present a simple yet accurate approach to modeling single sections of accumulation and non-accumulation conveyors with random load spacing and no branching or merging. In the case of non-accumulation conveyors, the approach recommended is a "follow-the-leader" approach in which all of the focus of attention is given to the leading part on the conveyor. When a new part becomes the leader, the remaining time needed for it to reach the end can be computed using the following four-step procedure:

1. Start with the minimum travel time needed for an object to move from its original entry position to its exit position.
2. Subtract the time it has already spent on the conveyor.
3. Add any conveyor halt time already experienced by the object as of the time it becomes the new leader.
4. As the simulation proceeds, add any additional conveyor halt time that the new leader experiences before it reaches the exit.

In the case of a simple accumulating conveyor on which parts continue if it comes to a halt, the approach is more simple. Parts take the minimum travel time after which they may be considered as being available for output. This will work when, as is the case in most instances, parts that are queued on the conveyor will not be pulled from the queue any faster than the time for one part to travel the distance of one queue position.

While these two simplified approaches may be adequate for rough modeling of single conveyor sections in which parts enter at one end and exit at the other, it is highly inadequate for a conveyor system in which parts are continuously branching and merging onto conveyor sections. This type of system requires continuous tracking which, in many discrete oriented simulation languages, tends to be impractical and gets very complex.

Modeling Conveyor Networks

A conveyor network consists of two or more sections that are connected together to enable part merging, diverting, and recirculating (see Figure 9.1). In such instances, a conveyor may have one or more entry points and one or more exit points. Furthermore, part entry and exit may not always be located at extreme opposite ends of the conveyor. You will remember that conveyor networks require an escort memory or computerized load control system to monitor and control part movement. Regardless of whether an escort memory system or a computerized load control system is used, the same decisions need to be made at each control point depending on the destination. Careful coordination is required to deliver the right part to the right station and to avoid part collisions at intersections.

Figure 9.1 Example of a Conveyor System with Loop, Merges, and Branches. *Filled and capped bottles travel up from the bottling area to packaging and shipping at the Molson O'Keefe Brewery in Montreal.*

Conveyor networks may provide alternative routes to various destinations. Part movement time on a conveyor is often only determinable by modeling the combined effect of the conveying speed, length of the conveyor, load spacing requirements, connecting conveyor logic and any conveyor transfer times. The traffic and conveyor dynamics ultimately determine the move time.

When modeling conveyor networks, it is imperative that the simulation product used provides the necessary constructs for modeling the network. An intelligent set of conveyor constructs will take defined conveyor sections along with their specified linkages, and automatically piece together the entire network, putting in load sensors for controlling traffic and load transfers from one section to another. This takes most of the difficulty out of modeling the conveyor. It does not, however, relieve the modeler from understanding how the conveyor dynamics work.

INDUSTRIAL VEHICLES

Industrial vehicles include all push or powered carts and vehicles that generally have free movement (see Figure 9.2). Powered vehicles such as lift trucks are usually utilized for medium distance movement of batched parts in a container or pallet. For short moves, manual or semi-powered carts are useful. Single load transporters are capable of moving only one load at a time from one location to another. Such devices are fairly straightforward to model because they involve only a single source and a single destination for each move. Multiple load transporters, on the other hand, can move more than one load at a time from one or more sources to one or more destinations. Defining the capacity and operation for multiple load transporters can be extremely difficult since there are special rules defining when to retrieve additional loads and when to deliver the loads already on board.

Figure 9.2 An Industrial Vehicle.

Efficient utilization of warehouse space is critical to managing inventory. Here, the cube of a warehouse is fully-utilized by stacking multiple units of the same product.
Photograph provided by permission from GATX Logistics.

One example of a multiple load transporter is a picking transporter which may be simply a person pushing a cart through a part storage area, loading several different part types onto the cart. These parts may then be delivered to a single destination perhaps to fill a customer order or to several destinations for the replenishment of stock along a production line. Towing vehicles are another example of this type of transporter. A towing vehicle pulls a train of carts behind it hauling several loads at once. Towing vehicles may have a single pickup and drop point for each move, similar to a unit load transporter, however they are also capable of making several stops, picking up multiple loads and then delivering them to several remote locations. Such an operation can become extremely complex to incorporate into a general modeling language.

Modeling Industrial Vehicles

Modeling an industrial vehicle involves modeling a resource that moves along a path network. Paths are typically open aisles in which bi-directional movement is possible and passing is permitted. Deployment strategies (work searches, idle vehicle parking, etc.) must be capable of being incorporated into the model, similar to that of an AGVS described later.

Also, industrial vehicles are generally human operated: they take breaks and may only be available during certain shifts. And, unless the vehicle is motorized, there are seldom failures associated with industrial vehicles such as push carts.

Performance measures

- Vehicle utilization
- Response time
- Move rate capability

Decision variables

- Number of vehicles
- Load pickup sequence
- Empty vehicle positioning

Questions to be answered

- What is the required number of vehicles to handle the required activity?
- What is the best deployment of vehicles to maximize utilization?
- What is the best deployment of empty vehicles to minimize response time?

AUTOMATED STORAGE/RETRIEVAL SYSTEMS (AS/RS)

An automated storage/retrieval system (AS/RS) is "a combination of equipment and controls which handles, stores, and retrieves materials with precision, accuracy, and speed under a defined degree of automation" (The Material Handling Institute 1977). The goal of an AS/RS is to provide random, high density storage with quick load access, all under computer control.

An AS/RS is characterized by one or more aisles of rack storage having one or more storage/retrieval (S/R) machines (sometimes called vertical or stacker cranes) that store and retrieve material into and out of a rack storage system. Material is picked up for storage by the S/R machine at an input (pickup) station. Retrieved material is delivered by the S/R machine to an output (deposit) station for take away. Usually there is one S/R machine per aisle, however, there may also be one S/R machine assigned to two or more aisles or perhaps even two S/R machines assigned to a single aisle. For higher storage volume, some AS/RSs utilize double deep storage which may require load shuffling to access loads. Where even higher storage density is required and longer term storage is needed, multidepth or deep lane storage systems are used. AS/RSs that use pallet storage racks are usually referred to as unit load systems, while bin storage rack systems are known as miniload systems.

The throughput capacity of an AS/RS is a function of the rack configuration. Throughput is measured in terms of how many single or dual cycles can be performed per hour. A single cycle is measured as the average time required to pick up a load at the pickup station, store the load in a rack location and return to the pickup station. A dual cycle is the time required to pick up a load at the input station, store the load in a rack location, retrieve a load from another rack location and deliver the load to the output station. Obviously, there is considerable variation in cycle times due to the number of different possible rack locations that can be accessed. If the AS/RS is a stand-alone system with no critical interface with another system, average times are adequate for designing the system. If, however, the system interfaces with other systems such as front end or remote picking stations, it is important to take into account the variability in cycle times.

Configuring an AS/RS. Configuring an AS/RS is primarily dependent on the storage/retrieval or throughput activity that is required and the amount of storage space required. Other considerations are building constraints in length, width, and height (especially if it is an existing facility); crane height limitations; and cost factors (a square building is less costly than a rectangular one).

Configuring an AS/RS is essentially an iterative process which converges on a solution that meets both total storage requirements and total throughput requirements. Once the number of rack locations are calculated based on storage requirements, an estimate of the number of aisles is made based on throughput requirements. This usually assumes a typical throughput rate of about 20 dual cycles per hour and assumes that one SR machine is assigned to an aisle. If, for example, the required throughput is 100 loads in and 100 loads out per hour, an initial configuration would probably be made showing five aisles. If, however, it was shown using a AS/RS simulator or other analytical technique that the throughput capacity was only 15 cycles per hour due to the large number of rack openings required, the number of aisles may be increased to seven or eight and analyzed again. This process is repeated until a solution is found.

Occasionally the throughput requirement is extremely low relative to the number of storage locations in the system. In such situations, rather than have one or two enormously long aisles, a man-aboard system will be used in which operator-driven vehicles will be shared among several aisles or perhaps a double deep storage rack will be used.

Computerized cycle time calculations. It is often felt that it is necessary to simulate an AS/RS in order to determine throughput capacity. This is because the S/R machine accesses random bins or rack locations, making the cycle time extremely difficult to calculate. The problem is further compounded by mixed single (pickup and store or retrieve and deposit) and dual cycles (a pickup, storage, retrieval, and deposit). Activity zoning, in which items are stored in assigned zones based on frequency of use, also complicates cycle time calculations. The easiest way to accurately determine the cycle time for an AS/RS is by using a computer to enumerate the possible movements of the S/R machine from the pickup and deposit (P&D) stand to every rack or bin location. This produces an empirical distribution for the single cycle time. For dual cycle times, an intermediate cycle time must be determined which is the time to go from any location to any other location. For a rack 10 tiers high and 40 bays long, this can be 400 x 400 or 160,000 calculations! Because of the large number of calculations, sometimes a large sample size is used to develop the distribution. Most suppliers of automated storage/retrieval systems have computer programs for calculating cycle times that can be generated based on a defined configuration in a matter of minutes.

Analytical cycle time calculations. Analytical solutions have been derived for calculating system throughput based on a given aisle configuration. Such solutions often rely on simplified assumptions about the operation of the system. Bozer and White (1984) for example, derive an equation for estimating single and dual cycle times assuming (1) randomized storage, (2) equal rack opening sizes, (3) P&D location at the base level on one end, (4) constant horizontal and vertical speeds, and (5) simultaneous horizontal and vertical rack movement. In actual practice, rack openings are seldom of equal size and horizontal and vertical accelerations can have a significant influence on throughput capacity.

While analytical solutions to throughput estimation may provide a rough approximation for simple configurations, there are other configurations that

Figure 9.3 The AS/RS at Bloomington-Normal Seating Company.
The AS/RS operates on a first-in, first-out basis. Complete finished inventory turns over once per day—approximately 1,400 storage and retrieval transaction with an accuracy of 99.99 percent.

become extremely difficult to estimate. In addition, there are control strategies that may improve throughput rate such as retrieving the load on the order list that is closest to the load just stored or storing a load in an opening that is closest to the next load to be retrieved. Finally, analytical solutions do not provide a distribution of cycle times but merely a single expected time which is inadequate for analyzing an AS/RS that interfaces with other systems.

AS/RS with picking operations. While some AS/RS (especially unit-load systems) have loads that are not captive to the system, many systems (particularly mini-load systems) deliver bins or pallets either to the end of the aisle or to a remote area where material is picked from the bin or pallet and then it is returned for storage. Remote picking is usually achieved by linking a conveyor system to the AS/RS where loads are delivered to remote picking stations. In this way, containers stored in any aisle can be delivered to any workstation. This permits entire orders to be picked at a single stations and eliminates the two-step process of picking followed by order consolidation.

Where picking takes place, an important goal is to achieve the highest productivity from both the AS/RS and the picker. Both are expensive resources and it is undesirable to have either one waiting on the other.

Modeling AS/RS

Simulation of AS/RS has been a popular area of application for simulation. Ford developed a simulator called GENAWS as early as 1972. IBM developed one in 1973 called the IBM Warehouse Simulator. Currently, most simulation of AS/RS is performed by vendors of the systems. Typical inputs for precision modeling of single deep AS/RS with each aisle having a captive S/R machine include:

- Number of aisles
- Number of S/R machines
- Rack configuration (bays and tiers)
- Bay or column width
- Tier or row height
- Input point(s)
- Output point(s)
- Zone boundaries and activity profile if activity zoning is utilized
- SR machine speed and acceleration/deceleration
- Pickup and deposit times
- Downtime and repair time characteristics

At a simple level, an AS/RS move time may be modeled by taking a time from a probability distribution that approximates the time to store or retrieve a load. More precise modeling incorporates the actual crane (horizontal) and lift (vertical) speeds. Each movement usually has a different speed and distance to travel which means that movement along one axis is complete before movement along the other axis begins. From a modeling standpoint, it is usually only necessary to calculate and model the longest move time.

In modeling an AS/RS, the storage capacity is usually not a consideration and the actual inventory of the system is not modeled. It would require lots of overhead to model the complete inventory in a rack with 60,000 pallet locations. Since it is primarily only the activity that is of interest in the simulation, actual inventory is ignored. In fact, it is usually not even necessary to model specific stock keeping units (SKU) being stored or retrieved, but only distinguish between load type insofar as it affects routing and subsequent operations.

Performance measures

- S/R machine utilization
- Response time
- Throughput capability

Decision variables

- Number of aisles
- Storage and retrieval sequences
- Empty SR machine positioning

Questions to be answered

- What is the required number of aisles to handle the required activity?
- What is the best sequence of stores and retrievals to maximize throughput?
- What is the best stationing of empty S/R machines to minimize response time?

CAROUSELS

One of the common classes of storage and retrieval systems are carousels. A carousel storage system consists of a collection of bins that revolve in either a horizontal or vertical direction. The typical front-end activity on a carousel is order picking. If carousels are used for WIP storage, bins might enter and leave the carousel. The average time to access a bin in a carousel is usually equal to one-half of the revolution time. If bi-directional movement is implemented, average access time is reduced to one-quarter of the revolution time. These times can be further reduced if special storing or ordering schemes are used that minimize access time.

A variation of the carousel is a rotary rack consisting of independently operating bin tiers. Rotary racks provide the added advantage of being able to position another bin while the operator is picking out of a bin from a different tier.

Modeling Carousels

Carousels are easily modeled by defining an appropriate response time representing the time for the carousel to bring a bin into position for picking. In addition to response times, carousels may have capacity considerations. The contents may even have an affect on response time, especially if the carousel is used to store multiple bins of the same item such as WIP storage. Unlike large AS/RS, storage capacity may be an important issue in modeling carousel systems.

AUTOMATIC GUIDED VEHICLE SYSTEMS (AGVS)

An automatic guided vehicle system (AGVS), is a path network along which computer-controlled, driverless vehicles transport loads. AGV systems are usually used for medium activity over medium distances. If parts are large, they are moved individually, otherwise parts are typically consolidated into a container or onto a pallet. Operationally, AGVS systems are more flexible than conveyors which provide fixed-path, fixed-point pickup and delivery. However, they cannot handle the high activity rates of conveyors. On the other extreme, AGVS systems are not as flexible as industrial vehicles that provide open path, any-point pickup and delivery. However, they can handle higher activity rates and eliminate the need for operators. AGVs provide a flexible, yet defined path of movement and allow pickup and delivery between any one of many fixed points except for where certain points have been specifically designated as pickup or drop off points.

AGVs are controlled by either a central computer, on-board computers, or a combination of the two. AGVs with on-board microprocessors are capable of having maps of the system in their memories. They can direct their own routing and blocking and can communicate with other vehicles if necessary. AGVs typically follow along an inductive guide path which is a wire buried a few inches beneath the floor. Travel is generally unidirectional to avoid traffic jams although occasionally a path might be bi-directional.

The AGVS control system must make such decisions as what vehicle to dispatch to pick up a load, what route to take to get to the pickup point, what to do if the vehicle gets blocked by another vehicle, what to do with idle vehicles and how to avoid collisions at intersections.

AGVs have acceleration and deceleration with speeds that often vary depending on whether vehicles are full or empty and whether they are on a turn or a stretch. There is also a time associated with loading and unloading parts. Furthermore, loaded vehicles may stop at points along the path for operations to be performed while the load is on-board the vehicle. One or more vehicles may be in a system and more than one system may share common path segments. AGVs stop periodically for battery recharging and may experience unscheduled failures.

One of the modeling requirements of AGVs is to accurately describe the method for controlling traffic. This is usually accomplished in one of two ways:

1. Zone blocking
2. On-board vehicle sensing

Zone blocking is the most common method of traffic control and involves placing control points along the guide path. Each control point usually allows only one vehicle to access it at any one time, thus blocking any other vehicle from entering any segment of the path connected to that point. Once the vehicle leaves a control point to travel to the next point on the path, any vehicle waiting for access to the freed control point can resume travel.

On-board vehicle sensing works by having a sensor on-board the vehicle that detects the presence of a vehicle ahead of it and stops until it detects that the vehicle ahead of it has moved.

Designing an AGVS

When designing an AGV system, the first step is to identify all of the pickup and drop off points. Next, the path should be laid out. In laying out the guide path, several principles of good design should be followed (Askin and Standridge 1993):

- Travel should be unidirectional in the form of one or more recirculating loops unless traffic is very light. This avoids deadlock in traffic.
- Pickup stations should be downstream of drop off stations. This enables an AGV to pick up a load, if available, whenever a load is dropped off.
- Pickup and deposit stations should generally be placed on spurs unless traffic is extremely light. This prevents AGVs from being blocked.
- Crossover paths should be used occasionally if measurable reductions in travel times can be achieved.

Once the path has been configured, a rough estimate is made of the number of AGVs required. This is easily done by looking at the total travel time required per hour, dividing it by the number of minutes in an hour (60), and multiplying it by one plus a traffic allowance factor as shown below:

$$\text{AGV's} = \frac{\text{total AGV minutes}}{\text{60 minutes/hr}} \times (1 + \text{traffic allowance})$$

The traffic allowance is a percentage factored into the vehicle requirements to reflect delays caused by waiting at intersections, rerouting due to congestion, etc. This factor is a function of both poor path layout and quantity of vehicles and may vary anywhere from zero in the case of a single vehicle system to around 15 percent in congested layouts with numerous vehicles (Fitzgerald 1985).

The total vehicle minutes is determined by looking at the average required number of pickups and deposits per hour, and converting it into the number of minutes required for one vehicle to perform these pickups and dropoffs. It is equal to the sum of the total empty move time, the total full move time and the total pickup and deposit time for every delivery. Assuming *I* pickup points and *J* drop off points, the total vehicle time can be estimated by using the following equation:

Total vehicle minutes =

$$\sum_{i=1}^{I}\sum_{j=1}^{J}\left[2 \text{ X } \frac{\text{Distance}_{ij}}{\text{Avg. speed}_{ij}} + \text{pickup}_i + \text{deposit}_j\right] \text{ X deliveries}_{ij}$$

The expression

$$2 \text{ X } \frac{\text{Distance}_{ij}}{\text{Avg. speed}_{ij}}$$

implies that the time required to travel to the point of pickup (empty travel time) is the same as the time required to travel to the point of deposit (full travel time). This assumption provides only an estimate of the time required to travel to a point of pickup since it is uncertain where a vehicle will be coming from. In most cases, this should be a conservative estimate since vehicles usually follow a work search routine in which the closest loads are picked up first, and vehicles frequently travel faster when empty than when full. A more accurate way of calculating empty load travel for complex systems is to use a compound-weighted-averaging technique that considers all possible empty moves together with their probabilities (Fitzgerald 1985).

Managing an AGVS

Since a multi-vehicle AGVS can be quite complex and expensive, it is imperative that the system is well managed to achieve the utilization and level of productivity desired. Part of the management problem is determining the best deployment strategy or dispatching rules used. Some of these operating strategies are discussed below (Adams 1985).

AGV selection rules. When a load is ready to be moved by an AGV and more than one AGV is available for executing the move, a decision must be made as to which vehicle to use. The most common selection rule is "closest rule"—the nearest available vehicle is selected. This rule is intended to minimize empty vehicle travel time as well as to minimize the part waiting time. Other less frequently used vehicle selection rules include:

- Longest idle vehicle
- Least utilized vehicle

Work search rules. If a vehicle becomes available and two or more parts are waiting to be moved, a decision must be made as to which part should be moved first. There are a number of rules that are used to make this decision, each of which can be effective depending upon the production objective. Common rules used for dispatching an available vehicle include the following:

- Longest waiting load
- Closest waiting load
- Highest priority load
- Most loads waiting at a location

Vehicle parking rules. If an AGV delivers a part and no other parts are waiting for pickup, a decision must be made relative to the deployment of the AGV. For example, the vehicle can remain where it is or it can be sent to a more strategic location where it is likely to be needed next. If several AGVs are idle, it may be desirable to have a prioritized list for a vehicle to follow for alternative parking preferences.

Work zoning. In some cases, it may be desirable to keep a vehicle captive to a particular area of production and not allow it to leave this area unless it has work to deliver elsewhere. In this case, the transporter must be given a zone boundary within which it is capable of operating.

Modeling an AGVS

Modeling an AGVS is very similar to modeling an industrial vehicle (which it is in a sense) except that the operation is more controlled and there exists less freedom of movement. Paths are generally unidirectional and no vehicle passing is allowed.

One of the challenges in modeling an AGVS in the past has been finding the shortest routes between any two stops in a system. Current state-of-the-art simulation software provides built-in capability to automatically determine the shortest routes between points in a complex network.

Performance measures

- Resource utilization
- Load movement rate
- Response time

In addition to the obvious purpose of simulating an AGVS to find out if the number of vehicles is sufficient or excessive, simulation can also be used to determine the following:

Decision variables

- Number of vehicles
- Work search rules
- Park search rules
- Placement of crossover and bypass paths

Questions to be answered

- What is the best path layout to minimize travel time?
- Where are the potential bottleneck areas?
- How many vehicles are needed to meet activity requirements?
- What are the best scheduled maintenance/recharging strategies?
- Which task assignment rules maximize vehicle utilization?
- What is the best idle vehicle deployment that minimizes response time?
- Is there any possibility of deadlocks?
- Is there any possibility of collisions?

Other Cart-type Systems

In addition to AGVs that travel on a floor, there are other types of computer-controlled, driverless carts that travel on a single track. A shuttle, for example, is a cart on a track generally used to move parts between a main transport system and a workstation. Shuttles provide precision handling for automated interfacing. These carts are self-propelled (either by an on-board energy source or through a power pickup) and follow paths that are defined either by rails or by an electronic signal. They also exhibit controlled acceleration/deceleration, variable speed control, and precise positioning for automated interfacing. Because of the speed control, carts can achieve high speeds while at the same time avoiding sudden starts and stops, cutting transit time to a minimum.

Modeling Shuttle Carts

Modeling shuttle cart systems that have one cart per track is fairly straightforward and is identical to modeling an industrial vehicle. It becomes much trickier, however, when multiple carts operate along a single bi-directional track. In such systems, the simulation must be able to model the situations where one cart must back out of the way of another cart. This same issue applies to cranes.

CRANES AND HOISTS

Cranes are floor, ceiling, or wall-mounted mechanical devices generally used for short to medium distances, discrete movement of material. Besides bridge and gantry cranes, most other types of cranes and hoists can be modeled as an industrial vehicle. Since a gantry crane is just a simple type of bridge crane having legs, our discussion here will be limited to bridge cranes.

Bridge Cranes

A bridge crane consists of a beam that bridges a bay (wall to wall). This beam or bridge moves on two tracks mounted on either wall. A retracting hoist unit travels back and forth under the bridge, riding on tracks. The bridge also travels on a track providing a total of three axes of motion (gantry cranes have similar movement, only the runway is floor mounted).

Crane Management

Managing crane movement requires an understanding of the priority of loads to be moved as well as the move characteristics of the crane. One must find the optimum balance of providing adequate response time to high priority moves while maximizing the utilization of the crane. Crane utilization is maximized when drop offs are always combined with a nearby pickup. Crane management becomes more complicated when (as is often the case) two or more cranes operate in the same bay. Pritsker (1986) identifies four typical rules derived from practical experience for managing bridge cranes sharing the same runway:

1. Cranes moving to drop off points have priority over cranes moving to pickup points.
2. If two cranes are performing the same type of function, the crane that was assigned its task earliest is given priority.
3. To break ties on one and two above, the crane closest to its drop off point is given priority.
4. Idle cranes are moved out of the way of cranes that have been given assignments.

These rules tend to minimize the waiting time of loaded cranes.

Modeling Bridge Cranes

Modeling a bridge crane requires much of the same kind of input as when modeling an AS/RS. Both cases deal with cranes having multiple axes of travel. In addition, every place a crane interfaces with pickup and deposit stations must be identified.

Performance measures

- Crane utilization
- Load movement rate
- Response time
- Percentage of time blocked by another crane

Decision variables

- Work search rules
- Park search rules
- Multiple crane priority rules

Questions to be answered

- Which task assignment rules maximize crane utilization?
- What idle crane, parking strategy minimizes response time?
- How much time are cranes blocked in a multi-crane system?

ROBOTS

Robots are programmable, multifunctional manipulators used for handling material or to manipulate a tool such as a welder for processing material. Robots are often classified by the type of coordinate system on which they are based. Typical coordinate systems are as follows:

- Cylindrical
- Cartesian
- Revolute

Choice of robot coordinate systems depends on the application. Cylindrical or polar coordinate robots are generally more appropriate for machine loading. Cartesian coordinate machines are easier to equip with tactile sensors for assembly work. Revolute or anthropomorphic coordinate robots have the most degrees of freedom and are especially suited for use as the processing tool such as welding or painting (FMS Handbook, Vol II, 1983). Since cartesian or gantry robots can be easily modeled as cranes, we will focus on cylindrical and revolute gantry robots. When used for handling, cylindrical or revolute robots are generally used to handle a medium level of movement activity over very short distances, usually to perform pick-and-place or load/unload functions. Robots generally move parts individually rather than in a consolidated load.

One of the applications of simulation is in designing the cell control logic for a robotic work cell. A robotic work cell may be a machining, assembly, inspection or a combination cell. Robotic cells are characterized by a robot with three to five degrees of freedom surrounded by workstations. The work station is fed parts by an input conveyor or other accumulation device and parts exit from the cell on a similar device. Each workstation usually has one or more buffer positions to which parts are brought if the workstation is busy. Like all cellular manufacturing, a robotic cell usually handles more than one part type and each part type may not have to be routed through the same sequence of workstations. In addition to part handling, the robot may be required to handle tooling or fixtures.

Robot Control

Robotic cells are controlled by algorithms that determine the sequence of activities for the robot to perform. This control is developed off line using algorithms for selecting what activity to perform. The sequence of activities performed by the robot can be either a fixed sequence of moves that is continually repeated or it can be a dynamically determined sequence of moves depending on the status of the cell. For fixed sequence logic, the robot control software consists of a sequence of commands that are executed sequentially. Dynamic sequence selection is programmed using "if-then" statements followed by branches depending on the truth value of the conditional statement. System variables must also be included in the system to keep track of system status, or else the robot polls the sensors of the system each time a status check is needed.

In developing an algorithm for a cell controller to make these decisions, the engineer must first decide if the control logic is going to be with respect to the robot, the part or the workstation. In a part-oriented system, the events associated with the part determine when decisions are made. When a part finishes an operation, a routing selection is made and then the robot is requested. In a robot-oriented cell, the robot drives the decisions and actions so that when a part completes an operation, it merely sends a signal or sets a status indicator showing that it is finished. When the robot becomes available, it sees that part is waiting and, at that moment, the routing decision for the part is made.

Modeling Robots

In modeling robots, it is sometimes presumed that the kinematics of the robot need to be simulated. Kinematic simulation is a special type of robot simulation used to do cycle time analysis and off-line programming. For discrete-event simulation, it is only necessary to know the move times from every pickup point to every deposit point, which usually are not very many. This makes modeling of a robot no more difficult than modeling a simple resource. The advantage to having a specific robot construct in a simulation product is primarily for providing the graphic animation of the robot.

Performance measures

- Robot utilization
- Response time

Decision variables

- Pickup sequence
- Idle robot positioning
- Robot scheduling algorithm

Questions to be answered

- What priority of movements results in the highest productivity?
- Where is the best position for the robot after completing each particular drop off?

SUMMARY

Whether configuring a material handling system or developing the control software, simulation is an extremely beneficial tool. Material handling systems can be one of the most difficult systems to model using simulation because of their complexity. We recommend that you simplify wherever possible when modeling material handling systems, yet do not overlook the issues involved.

Chapter 10
Modeling Service Systems

"No matter what line you move to, the other line always moves faster."

Unknown

INTRODUCTION

A service system is a processing system in which one or more services are provided to customers. Entities (customers, patients, paperwork, etc.) are routed through a series of processing areas (checkin, order, service, payment, etc.) where resources (service agents, doctors, cashiers, etc.) provide some service. Service systems exhibit unique characteristics that are not found in manufacturing systems. Sasser, Olsen, and Wyckoff (1978) summarize the distinct characteristics of service systems in four ways:

1. Services are *intangible*; they are not things.
2. Services are *perishable*; they cannot be inventoried.
3. Services provide *heterogeneous output*.
4. Services involve *simultaneous production and consumption*.

These distinctions pose great challenges for service systems design and management, particularly in the areas of facility layout, process design, equipment selection, and staffing. Having discussed general modeling procedures common to both manufacturing and service system simulation in Chapter 7, and specific modeling procedures unique to manufacturing systems in Chapter 8, this chapter discusses design and operating considerations that are more specific to service systems. Methodologies are presented for modeling different types of service systems.

APPLICATIONS OF SIMULATION IN SERVICE SYSTEMS

The use of simulation in service industries has been relatively limited in the past, despite the many areas of application where simulation has proven beneficial: healthcare services (hospitals, clinics, etc.), food services (restaurants, cafeterias, etc.), and financial services (banks, credit unions, etc.) to name a few. In healthcare services, Zilm et al. (1976) studied the impact of staffing on utilization and cost. Hancock et al. (1978) employed simulation to determine the optimum number of beds a facility would need to meet patient demand. Iskander and Carter et al. (1991) developed a model to evaluate alternative facility layouts for an ambulatory care center. In food services, Aran and Kang (1987) designed a model to determine the optimal seating configuration for a fast food restaurant. Kharwat et al. (1991) used simulation to examine restaurant and delivery operations relative to staffing levels, equipment layout, workflow, customer service, and capacity. Successful applications like these set the stage for future simulation studies in the service sector.

Even within manufacturing industries themselves there are business or support activities that are similar to those found in traditional service industries. Edward J. Kane of IBM observed (Harrington 1991):

> Just taking a customer order, moving it through the plant, distributing these requirements out to the manufacturing floor—that activity alone has thirty sub-process steps to it. Accounts receivable has over twenty process steps. Information processing is a whole discipline in itself, with many challenging processes integrated into a single total activity. Obviously, we do manage some very complex processes separate from the manufacturing floor itself.

This entire realm of support processes presents a major area of application for simulation. Similar to the problem of dealing with excess inventory in manufacturing systems, customers, paperwork, and information often sit idle in service systems while waiting to be processed. In fact, the total waiting time may be as high as ninety-five percent of the total processing time.

The types of questions that simulation helps answer in service systems can be categorized as being either design related or management related. However, it is difficult to generalize the issues that can be addressed by simulation for every type of service system. A list of typical questions that can be addressed by simulation is provided below:

I. System Design Decisions

- What is the capacity of the service and waiting areas?
- What is the maximum throughput capability of the service system?
- What are the equipment requirements to meet service demand?
- How long does it take to service a customer?
- How long do customers have to wait before being serviced?

- Where should the service and waiting areas be located?
- How can workflow and customer flow be streamlined?
- What effect would automation have on reducing non-value-added time?

II. System Management Decisions

- What is the best way to schedule staffing?
- What is the best way to schedule appointments for customers?
- What is the best way to schedule carriers or vehicles?
- How should the resources be assigned to tasks?
- Which customers or tasks should be serviced first?
- What is the best way to schedule maintenance for equipment and facilities?
- What is the best way to deal with emergency situations such as equipment failure?

PERFORMANCE MEASURES

The ultimate objectives of a service business include maximizing profits and customer satisfaction. However, these measures for success are considered to be *external performance criteria*, because they are not completely determined by any single activity. Simulation modeling and analysis help evaluate those measures referred to as *internal performance criteria*. These are measures that are solely within the control of a given activity. Simulation performance measures are both quantitative and time-based and measure the efficiency and the effectiveness of a system configuration and operating logic. Examples of output measures are waiting times, hours to process an application, cost per transaction, and percentage of time spent correcting transaction errors. D.A. Collier (1994) uses the term "interlinking" to define the process of establishing quantitative, causal relationships such as these to external performance measures. He argues that interlinking can be a powerful strategic and competitive weapon. Below is a list of the typical internal performance measures that can be evaluated using simulation.

Customer service time. Customer service time is the overall time from when the customer enters the system to the time the customer leaves the system. The customer service time includes the sum of the following:

- Service time
- Move time
- Queue or waiting time

Customer waiting time. The time a customer waits before a particular service is provided. Reducing customer waiting time contributes significantly to reducing total customer service time. It also affects customer satisfaction. A customer may be in the system waiting for a number of reasons:

- Waiting for diagnosis
- Waiting to order
- Waiting for price quotation
- Waiting for next available service agent
- Waiting for personnel to become available
- Waiting to receive a product or service
- Waiting for document processing
- Waiting for checkout

Number of customers waiting. The number of customers waiting for a service at any given time. This measure is also referred to as number of customers in queue. Larger numbers of customers waiting result in increased waiting times and longer overall customer service times.

Resource utilization. The percent of the time a resource (a server or unit of equipment) is busy or in use as a percentage of the total scheduled time. A high resource utilization may indicate efficiency, but may also be an indication of bottlenecks or long service times.

Order processing time. The time from when a customer places an order to the time the product or service is received. This performance measure is particularly significant for those types of services where customers leave the service facility after placing the order (professional service or retail service for large items). Order processing time is different from the total customer service time because it does not include the time a customer may spend in the service system for other purposes such as evaluating the service or product or actually placing the order. Increases in order processing times could cause customers to cancel orders resulting in lost customers.

Abandonment rate. The percentage of the time customers leave the arrival queue or decide not to enter the service system. High values of this performance measure indicate insufficient capacity.

Reliability. The consistency with which a particular standard of performance can be met. This might be measured in terms of errors or upsets per thousand observations.

Cost. Cost is the total financial measure of providing the service. Since time is money, cost of providing a service should be considered along with the time to perform the service. Cost is usually measured in terms of dollars per customer or transaction.

MODELING CONSIDERATIONS

Service systems represent a class of processing systems where entities (customers, orders, work, etc.) are routed through a series of service stations and waiting areas. Although certain characteristics of service systems are similar to manufacturing systems, service systems have some very unique characteristics. The aspects of service systems that involve workflow processing (orders, paperwork, records, etc.) and product delivery are nearly identical to manufacturing and will not be repeated here. Those aspects of service systems that are most different from manufacturing systems are those involving customer processing. Many of the differences stem from the fact that in service systems, often both the entity being served and the resource performing the service are human. Humans have much more complex and unpredictable behavioral characteristics than parts and machines. These special characteristics and their implications on modeling are described below.

Entities are Capricious

System conditions cause humans to change their minds about a particular decision once it has been made. Customer reactions to dissatisfactory circumstances include *balking*, *jockeying*, and *reneging*. Balking occurs when a customer attempts to enter a queue, sees that it is full, and leaves. Jockeying is where a customer moves to another queue that is shorter in hopes of being served sooner. Reneging is where a customer enters a waiting line or area, gets tired of waiting, and leaves. Modeling these types of situations can become complex and require special modeling constructs or the use of programming logic to describe the behavior.

Entity Arrivals are Random and Follow Cyclical Patterns

Customers arrive to most service systems randomly according to a Poisson process (interarrival times are exponentially distributed). Additionally, the rate of arrivals often changes depending on the time of day, or day of the week. The pattern of arrival rate change usually repeats itself in a cyclical fashion on a daily, weekly, or sometimes monthly basis. Accurate modeling of these arrival patterns and cycles are essential to accurate analysis.

Resource Decisions are Complex

Typically, resource allocation and task selection decisions are made according to some general rule (first come, first served, etc.). In service systems, however, resources are intelligent and often make decisions based on more state-related criteria. An increase in the size of a waiting line at a cashier, for example, may prompt a new checkout lane to open. A change in the state of an entity (a patient in recovery requiring immediate assistance) may even cause a resource (a nurse or doctor) to interrupt its current task to service the entity. The flexibility to respond to state changes is made possible because the resources are humans who are capable of making more complex decisions than machines. Modeling the complex behavior of human resources often requires the use of if-then logic to define the decision rules that are used.

Resource Work Pace is Variable

Another characteristic of resources that are human is that work pace tends to vary with time of day or work condition. A change in the state of a queue (number of customers in line) may cause a resource (cashier) to work faster thereby reducing processing time. A change in the state of the entity (length of waiting time) may cause a resource to work faster to complete the professional service. A change in the state of the resource (fatigue or learning curve) may change the service time (slower in the case of fatigue, faster in the case of a learning curve). To model this variable behavior, tests must be continually made on state variables in the system in order to link resource behavior to the system state.

Processing Requirements are Highly Variable

Service processes vary considerably due to the nature of the process as well as the fact that the entity and the server are both human. Consequently, processing times tend to be highly variable. From a modeling standpoint, processing times usually need to be expressed using some probability distribution such as a normal or beta distribution.

Services Have Both Front Room and Back Room Activities

In service systems, there often exists both front room and back room activities. In front room activities, customer service representatives meet with customers to take orders for a good or service. In the back room, the activities are carried out for producing the service or good. Once the service or good is produced, it is either brought to the customer who is waiting in the front room or, if the customer has gone, it is delivered to a remote site.

GENERAL SIMULATION PROCEDURES

Service systems are often analyzed using improper methods that are based on inaccurate assumptions. One mistake that is frequently made in planning service systems is using average rates and times to make design and management decisions. Frequently, such systems fail to meet expectations. The reason for this is that static and linear calculations ignore the effects of variability and interdependencies on system performance measures. For example, making a decision to staff a service system based on the average expected number of customers per hour does not take into account peak period demand for service. The system behavior during peak periods may be one of the most useful performance measures in designing a service facility. Simulation models can account for variation as well as system complexity and therefore provide the most accurate analysis of service system performance. However, even when simulating service systems, one must be careful to accurately represent the characteristics of the system and correctly interpret the output results.

Since service systems are nearly always in a state of transition, going from one activity level to another during different periods of the day or week, they rarely reach a steady-state condition. Consequently, we are frequently interested in analyzing the transient behavior of service systems. Questions such as "How long is the transient cycle?" or "How many replications to run?" become very important. Overall performance measures may not be as useful as the performance for each particular period of the transient cycle. An example where this is true is in the analysis of resource utilization statistics. In the types of service systems where arrival patterns and staff schedules are different over the activity cycle (day, week, etc.), the average utilization for the entire cycle is almost meaningless. It is much more informative to look at the resource utilization during different periods of the activity cycle.

Since most design and management decisions in service systems involve answering questions based on transient system conditions, it is important that the results of the simulation provide measures of transient behavior. Multiple replications should be run with statistics gathered and analyzed for different periods of the transient cycle. In an attempt to simplify a simulation model, sometimes there is a temptation to model only the peak period, which is often the period of greatest interest. What is overlooked is the fact that the state of the system prior to each period and the length of each period significantly impact the performance measures for any particular period, including the peak period.

SERVICE DECISION VARIABLES

Layout Decisions

One of the most important considerations in designing a service system is the layout of the facility. The layout must provide for convenient workflow and customer flow. Blackburn's observation at a large insurance company accurately describes the critical relationship between the layout and process flow (1992). He notes that "When the layout was superimposed on the (process) flowchart, the team observed that the paperwork flow crisscrossed two floors of a large office building." A good layout should be designed simultaneously with the design of the service process since they are so inextricably connected.

Automation Decisions

Automation of service processes presents one of the greatest opportunities for reducing the waiting time in the service process. However, the automation of a service process presents similar challenges to the automation of a manufacturing process. If automating a service process speeds up the process but does not minimize the overall processing time, it is not effective and may even be creating waste (large pools of waiting entities). Automation in the service sector can be applied to customer processing as well as to record and information processing. Other automation decisions that can be evaluated with simulation are the transportation of customers and resources.

Policy Decisions

Service managers are constantly faced with the problem of juggling resources and adapting to changing policies. In fact, a service system's agility can be considered a strategic advantage in competitive situations. For example, airlines occasionally offer special fares for limited times or a bank may decide to offer services on Saturdays. These types of policy decisions require the service providers to maintain service quality while operating under atypical conditions.

Workstation Design Decisions

The way in which workstations are designed can have a significant impact on customer satisfaction and processing efficiency. In some systems, for example, multiple servers have individual queues in front of them for customers to wait before being served. This can cause customers to be served out of order of arrival and results in jockeying and customer discontent. Other systems provide a single input queue which feeds multiple servers (queuing for a bank teller is usually designed this way). This ensures customers will be serviced in order of arrival. It may, however, cause some reneging if grouping all customers in a single queue creates the perception to an incoming customer that the waiting time is long.

Staffing Decisions

A major decision in nearly every service operation pertains to the level of staffing to meet customer demand. Understaffing can lead to excessive waiting times and lost or dissatisfied customers. Overstaffing can result in needless costs for resources that are inadequately utilized. Modeling staffing requirements is done by defining the pattern of incoming customers, specifying the servicing policies and procedures, and setting a trial staffing level.

After running the simulation, waiting times, abandonment counts and resource utilization rates can be evaluated to determine if the optimum conditions have been achieved. If results are unacceptable, either the incoming pattern, the service force or the servicing policies and procedures can be modified to run additional experiments.

Flow Control Decisions

Flow control is to service systems what production control is to manufacturing systems. Service system operations planners must decide how to allow customers, documents, etc. to flow through the system. Like manufacturing systems, customers and information may be pushed through the system or pulled through. By limiting queue capacities, a pull system can be achieved that reduces the total number of customers or items waiting in the system at any one time. It also reduces the average waiting time of customers or items and results in greater overall efficiency. Fast-food restaurants practice pull methods when they keep two or three food items (burgers, fries, etc.) queued in anticipation of upcoming orders. When the number of a particular item reaches a limit, no more of that item is prepared. When a server withdraws an item, that is the pull signal to the cook to replenish it with another one. The shorter the inventory of prepared food items, the more tightly linked the pull system is to customer demand. Excessive inventory results in deteriorating quality (cold food) and waste at the end of the day.

TYPES OF SERVICE SYSTEMS

There are several ways to classify service systems: type of service (financial, food, etc.) or the business purpose (profit or non-profit organizations). The classification of service systems provided here is based on operational characteristics and is adapted, in part, from the classification given by Schmenner (1994). These systems are as follows:

- Service factory
- Pure service shop
- Retail service stored
- Professional service
- Telephonic service
- Delivery service
- Transportation service

This chapter discusses each type of service system, looking at the performance measures, decision variables, and modeling considerations for each.

SERVICE FACTORY

Service factories are systems in which customers are provided services using equipment and facilities requiring low labor involvement. Consequently, labor costs are low while equipment and facility costs are high. Service factories usually have both front room and back room activities with total service being provided in a matter of minutes. Customization is done by selecting from a menu of options previously defined by the provider. Waiting time and service time are two primary factors in selecting the provider. Convenience of location is another important consideration. Customer commitment to the provider is low because there are usually alternative providers just as conveniently located.

Examples

Banks (branch operations), restaurants, copy centers, haircutters, and checkin counters of airlines, hotels, and car rental agencies.

Performance Measures

- Average service time by period
- Average waiting time by period
- Queue lengths by period
- Number of customers in the facility over time
- Abandonment rate during each period

Decision Variables

- Number of servers during each period
- Quantities of equipment
- Size of facilities (waiting area, parking)
- Arrival rate over the service cycle
- Hours of operation
- Hours of cleaning and maintenance
- Length of wait before reneging occurs
- Length of line before balking occurs
- How much shorter a neighboring line is before jockeying

Questions to be Answered

- How many of each type of equipment are required to meet customer demand?
- Which layout provides the most efficient customer flow and minimizes delays?
- Which resources can be shared to assist in peak times to minimize waiting time?

- How many shifts and service providers are needed to minimize costs?
- What procedures can be used (self-service, advance ordering) to minimize service time?

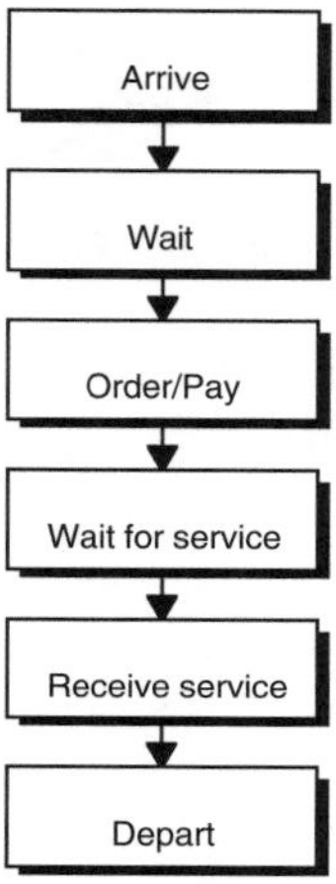

Arrive

Wait

Order/Pay

Wait for service

Receive service

Depart

Figure 10.1
Flow Diagram for Service Factory.

Model Representation

System element	*Model representation*
Customers, Orders	Customers and orders are the entities in the model.
Servers, Equipment	Servers and equipment can be modeled either as route locations or as resources. It is often appropriate to model equipment that is stationary, such as a copy machine, as a location since customers are routed to these items rather than having the items brought to the customer. Servers (personnel), on the other hand, are generally used at multiple locations. So, it is more appropriate to model servers as resources. An exception would be a resource dedicated to single location. In this case, the resource may be ignored. Preemption of resources (example, an unsatisfied customer preempting a server), scheduling of resources, shifts, learning curves, and priorities are important considerations when modeling resources.
Queues, Waiting areas	Queues and waiting areas are route locations where customers (entities) wait their turn to be serviced or where they wait to receive a good or service. When modeling queues, the modeler must specify the

physical capacity of the queue (how many customers can simultaneously wait in the queue), and the queuing rule, that is, which customer is serviced first (typically FIFO). Balking, reneging, and jockeying rules may also be defined with respect to a queue.

Arrivals

Typically, the interarrival times of customers or orders are exponential. Each arrival may contain either single or multiple entities. Arrivals are usually predictable based on time of day, day of week, etc.

Entity movement

Customers typically move through the system on their own, without an escort. Orders are moved from place to place by the service representatives, although sometimes orders are moved electronically. Entity movement times are relatively small compared to the actual service times and are sometimes simply added as part of the service time itself.

Processes

There are three major processing activities that are included in service shop models: order taking, the service activity, and receipt of the good or service. When a customer places an order, attributes of that order must be defined so that the order can be matched with the right customer at the point of pickup.

Routing

Customers are usually routed to one of several servers, usually depending on the length of the queue in front of the server. The type of service desired may also dictate which server to choose.

Simulation Procedure

Since service factories are transient behaving systems with daily or weekly cycle patterns, the simulation time should be either a day or a week. Multiple replications are needed to get an accurate estimate of performance measures for each phase of the cycle.

PURE SERVICE SHOP

In a pure service shop, service times are longer than for a service factory. Service customization is also greater. Customer needs must be identified before service can be provided. Customers may leave the location and return for pick up, to check on an order, make a payment, or for additional service at a later time. Price is often determined after the service is provided. Although front room activity times may be short, back room activity times may be long, typically measured in hours or days. The primary consideration is quality of service. Delivery time and price are of secondary importance. The customer's ability to describe the symptoms and possible service requirements are helpful in minimizing service and waiting times.

When customers arrive, they usually all go through some type of checkin activity. At this time, a record (paperwork or computer file) is generated for the customer, and a sequence of service or care is prescribed. The duration of the service or the type of resources required may change during the process of providing service because of a change in the status of the entity. After the service is provided, tests can be performed to ensure that the service is acceptable before releasing the entity from the facility. If the results are acceptable, the customer and the record are matched and the customer leaves the system.

Examples

Hospitals, repair shops (automobiles), equipment rental shops, banking (loan processing), Department of Motor Vehicles, Social Security offices, court rooms, and prisons.

Performance Measures

- Service time by period
- Waiting time by period
- Queue length by period
- Number of customers in the facility by period
- Resource utilization by period
- Throughput by period

Decision Variables

- Number of service providers
- Service times
- Size and location of facilities (waiting areas, service areas, parking, restrooms, etc.)
- Number and type of service and transportation equipment
- Capacity of the facility
- Sequencing of customers waiting for service
- Number of staff and shift schedules
- Hours of operation
- Maintenance schedules

Questions to be Answered

- Which layout provides the most convenient customer flow and minimizes delays?
- What is the peak capacity of the system?
- Which resources can be shared in peak times to minimize waiting time?
- How many shifts and service providers do we need to minimize costs?
- What procedures can be used (self diagnosis, advance checkin) to minimize service time?

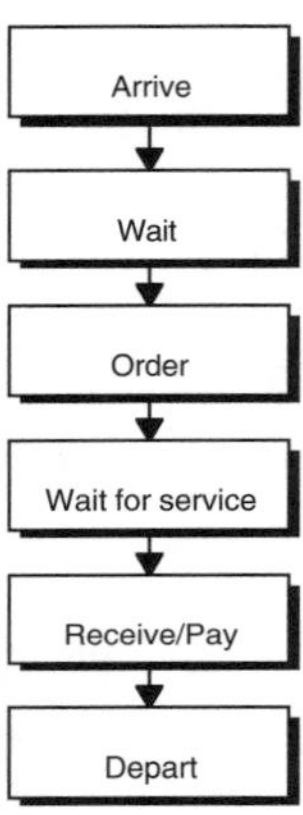

Figure 10.2
Flow Diagram for Pure Service Shop.

Model Representation

System element	*Model representation*
Customers, patients, records	Customers, patients, and records are represented by entities in the model. Customers may be classified into groups that possess similar characteristics or routings (patients may be classified based on DRGs). It is important to define attributes for customers (or classes of customers) and records to keep track of unique characteristics such as type, service time, and cost.
Service areas, rooms	The areas or rooms where customers receive service are the route locations in the model. They may be multi-capacity (accommodating multiple customers at any one time) or single capacity locations (only one customer occupies the location at a time).
Queues, waiting areas	Queuing areas are places where customers wait for service. When modeling these queuing locations, decision rules for queuing and location selection are important considerations. Balking, reneging and

jockeying rules may also be defined with respect to a queue.

Servers, Nurses, Doctors, Technicians, Equipment

These are resources that are needed to provide services to customers. They may be dedicated to a service area (equipment) or they may share service areas with other resources (doctors). Preemption of resources (a patient with a problem may require a nurse that is attending another patient), scheduling of resources, shifts, learning curves, and priorities are important considerations when modeling resources.

Arrivals

Demand may vary hourly, daily, weekly, monthly, or seasonally. Customer arrivals can be recorded and fit to distributions using time windows such as morning, afternoon, night, or similar periods. Accurate modeling of arrival distributions and cycles is essential to valid models.

Entity movement

Customers may move through the service shop on their own or with assistance from resources (technician, nurse). If customer movement is critical to the performance of the system, the paths that customers take should be modeled, including the movement characteristics of customers.

Processes

Processes include the following and may be performed by a single resource or multiple resources:

- Checkin
- Preliminary work
- Examination and diagnosis
- Treatment or repair
- Lab work, X-ray processing, etc.
- Consultation
- Checkout

Routing

Routing is usually dependent on the patient type or the types of services requested. This is usually done at a checkin point. At checkin, usually an expert resource evaluates the service needed by the customer and determines the type of service, estimated service time and cost. This diagnosis then dictates the routing of entities thereafter.

Decision rules — Decision rules include queuing rules, resource allocation rules, entity selection rules, priority rules, preemption, and room scheduling rules. Programming may be needed to model complex decision rules involving entity, location, or resources.

Simulation Procedure

Pure service activities are transient in nature with daily, weekly, or monthly cycle patterns. The simulation time should be a day, week, or month depending on the cycle. Multiple replications are needed to get an accurate estimate of performance measures for each period or phase of the cycle.

RETAIL SERVICE STORES

In retail services, the size of the facility is large in order to accommodate many customers at the same time. Customers are provided a large number of product options from which to choose. Retail services require a high degree of labor intensity but low degree of customization or interaction with the customer. Customers are influenced by price more so than service quality or delivery time. Customers are interested in convenient location, assistance with finding the products in the store, and quick checkout. Total service time is usually measured in minutes.

When customers arrive in a retail shop, they often get a cart and use that cart as a carrier throughout the purchasing process. Customers may need assistance from customer service representatives during the shopping process. Once the customer has obtained the merchandise, then he must get in line for the checkout process. For large items such as furniture or appliances, the customer may have to order and pay for the merchandise first. The delivery of the product may take place later.

Examples

Department stores, grocery stores, hardware stores, and convenience stores

Performance Measures

- Time waiting for assistance for each period
- Number waiting in checkout lines by period
- Time waiting for checkout by period
- Resource utilization by period
- Number of carts available over time

Decision Variables

- Number of servers
- Number of checkout stands
- Number of dock doors for delivery and pickup
- Number of carts
- Size of the facility
- Location of merchandise, carts, and service desks
- Shifts for cashiers and customer service representatives
- Replenishment frequency and quantity of inventory
- Hours of operation
- Maintenance schedules

Questions to be answered

- Which layout provides the most convenient customer flow and minimizes delays?
- Which resources can be shared to assist in peak times to minimize waiting time?
- How many shifts and service providers do we need to minimize costs?

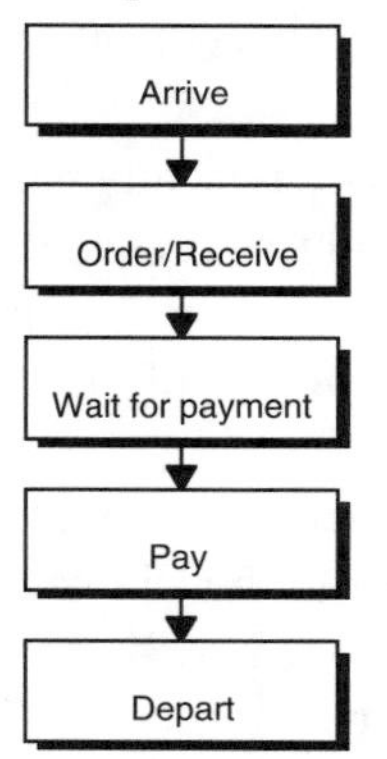

Figure 10.3
Flow Diagram for Retail Services Shop.

Model Representation

System element	*Model representation*
Customers, shopping carts	Customers and shopping carts are modeled as entities.
Queues, waiting areas	Queues in retail services mainly occur at customer checkout, or waiting for delivery of merchandise. When modeling these queuing locations, decision rules for queuing and location selection are important considerations. Jockeying and reneging are frequent occurrences in a retail store.

Entity movement	Customers move through the service shop on their own. Movement of customers within the retail shop is not of significant concern.
Checkout and customer service counters	These are service locations where customers pay or return merchandise. They may be multi-capacity or single capacity locations.
Cashiers, servers, managers	These are resources that provide services to customers in the retail shop. They may be dedicated to a service area or share service areas with other resources. Preemption of resources (a customer with a complaint or return may preempt a current customer), scheduling of resources, shifts, learning curves, and priorities are important considerations when modeling resources.
Arrivals	Customer arrivals may contain single or multiple customers. Accurate modeling of arrival distributions and cycles is essential.
Processes	Processes typically modeled include getting a cart and retrieving merchandise, requesting customer service, ordering, checkout, payment, credit approval, and replenishment of merchandise by resources.
Decision logic	The checkout process is perhaps the most difficult process to model in a retail shop. Therefore, programming code may be needed to model the complexity of checkout processes. The duration of the checkout process depends on the number of available servers, the customer's readiness for payment or credit, the number of items purchased by the customer, etc. During checkout, customers waiting in line exhibit behavior that is difficult to model, such as reneging or jockeying. These activities require continuous tracking of system state variables (number of customers in a particular queue) for decision making. Another dynamic element of a retail shop is the number servers in the checkout process. The checkout may be performed by a single resource or multiple resources depending on the number of customers in the queue.

Simulation Procedure

Retail service systems are usually non-steady-state systems that require a determination of the operation cycle and number of replications. The service cycle may vary from day to day so the simulation time is usually a week. Statistics should be gathered that reflect different periods in the cycle.

PROFESSIONAL SERVICES

Professional services are usually provided by a single person or a small group of experts in a particular field. The service is highly customized and provided by expensive resources. Duration of the service is long, extremely variable, and difficult to predict because the customer involvement during the process is highly variable.

Processing may be performed by a single resource or multiple resources. When the customer arrives, the first process is of diagnostic nature. Usually, an expert resource evaluates the service needed by the customer and determines what type of service, estimated service time, and cost. This diagnosis then dictates what resources are used to process the order. The duration of the service or the type of resources required may change during the process of providing service. This is usually a result of the customer's review of the work. After the service is provided, a final review with the customer may be done to make sure that the service is acceptable. If the results are acceptable, the customer and the record are matched and the customer leaves the system.

Examples

Auditing services, tax preparation, legal services, architectural services, construction services, and tailoring services.

Performance Measures

- Average service time
- Resource utilization
- Time spent doing rework

Decision Variables

- Number and type of service providers
- Staff schedules and shifts
- Project review times
- Hours of operation (overtime)

Questions to be Answered

- How many of each type of equipment are required to meet project deadline?

- Which resources can be shared to assist in making up lost time?
- How many shifts and service providers are needed to minimize costs?
- How can the available resources be scheduled to meet deadline?

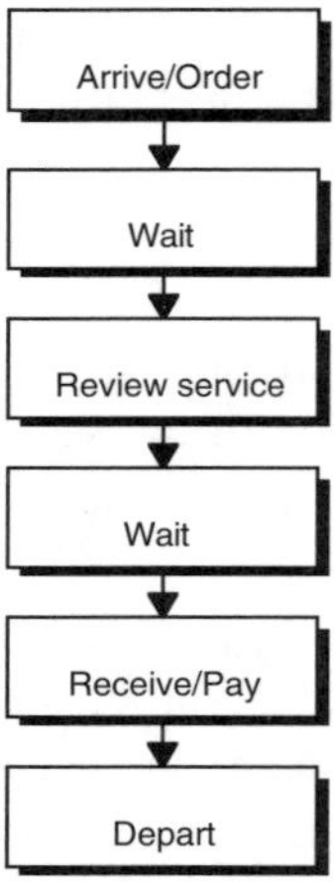

Arrive/Order → Wait → Review service → Wait → Receive/Pay → Depart

Figure 10.4
Flow Diagram for Professional Services.

Model Representation

System element	***Model representation***
Customers, orders	Customers and orders are modeled as entities. In a professional service shop, there are very few simultaneous customers or orders in the system. Attributes of orders such as due date and cost are important for gathering statistics for performance measures.
Arrivals	Customer arrivals are highly unpredictable or seasonal. Accurate modeling of arrival distributions and cycles is essential.
Queues	Physical queues in a professional service shop are almost non-existent because the customers do not wait for the service at the point of delivery. Queues may still be defined, however, to model the waiting state of customers.
Service areas	Usually there is only one service area in a professional shop. This is the location where the service is prepared and provided. In a professional service, the concern is with modeling the time over which the service is provided rather than the space in which it occurs.

Professionals, workers These are resources that provide professional services to customers. This is the single most important element in modeling a professional service. Typically, multiple resources may be working on the same order. Preemption of resources, scheduling of resources, shifts, learning curves, and fatigue are important considerations when modeling resources in a professional service shop.

Processes Processes to be included in the model include estimation, setup, use of multiple resources, service time, review, change of status, match customer with order, final review, and wait.

Routings Routings are quite simple when a customer enters, leaves a job request and leaves.

Decision logic Decisions are based on resource allocation rules, entity selection rules, priority rules, and preemption. Due to the high resource-intensive nature of a professional service shop, programming may be needed to model complex resource allocation and assignment rules. For example, one may wish to monitor the due date of the order and add resources based on delays.

Simulation Procedure

Professional services are sometimes seasonal but still usually reach steady-state behavior and therefore can be simulated using steady-state simulation methods.

TELEPHONIC SERVICES

Telephonic services or teleservicing are services provided over the telephone. They are unique from other services in that the service is provided without face-to-face contact with the customer. The service may be making reservations, catalogue ordering, or providing a customer support service. In a telephonic service system, issues to address include the following:

- *Overflow calls.* The caller receives a busy signal.
- *Reneges.* The customer gets in but hangs up after a certain amount of time if no assistance is received.
- *Redials.* A customer who hangs up or fails to get through calls again.

The most important criteria for measuring effectiveness is service time. The customer is simply interested in getting the service or ordering the product as quickly as possible. The customer's ability to communicate the need is critical to the service time.

Calls usually arrive in the incoming call queue and are serviced based on FIFO rule. Some advanced telephone systems allow routing of calls into multiple queues for quicker service. Processing of a call is done by a single resource. Duration of the service depends on the nature of the service. If the service is an ordering process, then the service time is short. If the service is a technical support process, then the service time may be long or the call may require a callback after some research.

Examples

Technical support services (hotlines) for software or hardware, mail order services, and airline and hotel reservations.

Performance Measures

- Service time by period
- Waiting time by period
- Abandonment rate by period

Decision Variables

- Number of operators
- Frequency of incoming calls by period
- Capacity of the phone system
- Staff schedule and shifts
- Call routing
- Hours of operation

Questions to be Answered

- How many shifts and service providers are needed to minimize costs?
- Which resources can be shared to assist in peak times to minimize waiting time?
- What automation technologies can be used to minimize service time?
- How can calls be routed to minimize waiting time?

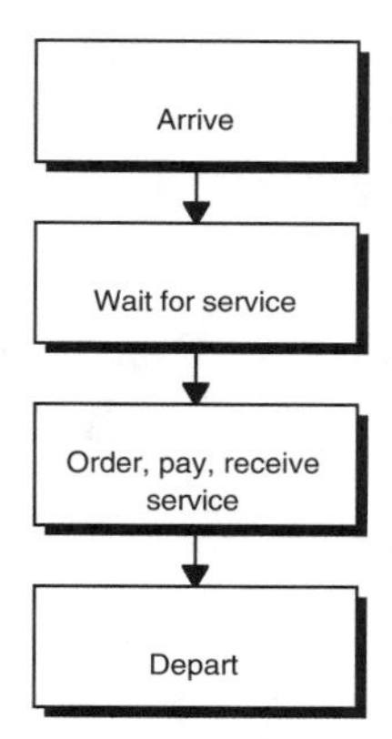

Figure 10.5
Flow Diagram for Telephonic Services

Model Representation

System element	*Model representation*
Calls	Telephone calls are modeled as entities. It is important to define attributes for customers and records in order to keep track of unique characteristics such as type and service time.
Queues	Queues should be used to model calls that are waiting. The queue length for incoming calls depends on the capacity of the phone system. When modeling queuing locations, decision rules for queuing and location selection are important considerations. Once a call is in the queue, the caller may hang up (reneging) after waiting for a while.
Customer service, representatives, computers	These are resources that provide services to customers over the phone. Because they are dedicated to specific telephones while providing the service, it is appropriate to model them as static resources or multi-capacity locations. Modeling shifts for the customer service representatives is an important consideration.
Arrivals	Although arrivals are usually exponentially distributed, the cycles for calls may be fairly predictable based on time of day, day of week, etc. Accurate modeling of arrival distributions and cycles is essential. Unlike other types of service systems, an arrival may not enter into the system if the phone lines are busy. This is an important aspect of modeling arrivals

in telephonic service shops because computing the number of calls that arrived but could not get in the queue can be helpful in determining the capacity of a phone system.

Processes

The most difficult aspect of modeling telephonic service shops is the queuing of incoming calls. Process modeling includes time to take an incoming call and make a routing determination, and time to provide each of the services.

Simulation Procedure

The situation being examined in the simulation is a transient one requiring a determination of the cycle period and number of replications. Of particular interest is capturing peak activities.

DELIVERY SERVICES

Delivery services involve the ordering, shipping and delivery of goods (raw materials or finished products) to points of use or sale. Customers may accept deliveries only within certain time schedules. In practice, there are often other constraints besides time windows. Certain sequences in deliveries may be inflexible. Customers are interested in convenient and fast delivery. If the products that are delivered are perishable or fragile goods, the quality of the products delivered is also important to the customer.

Deliveries begin with the preparation of the product and loading of the product on the delivery resources. Determination of the best routing decisions for drivers may depend on the number of customers waiting for the product or the proximity of customers waiting for the product.

Examples

Mail and package delivery, food delivery, flower delivery, and moving services

Performance Measures

- On-time delivery (or delivery within the required time window)
- Travel costs
- Total travel time
- Vehicle usage and carrying capacity utilization

Decision Variables

- Number of vehicles and drivers
- Facility size and layout

- Routing sequence
- Frequency of deliveries and delivery size
- Pickup loads from vendors, etc. on the backhaul or return to the warehouse where possible to avoid deadheading (i.e. driving empty).

Questions to be Answered

- How many of each vehicle are required to meet customer demand?
- Which delivery routes will maximize productivity?
- How many loads per vehicle maximizes the productivity?
- How many shifts and service providers are needed?
- What procedures can be used to minimize service or product selection time?

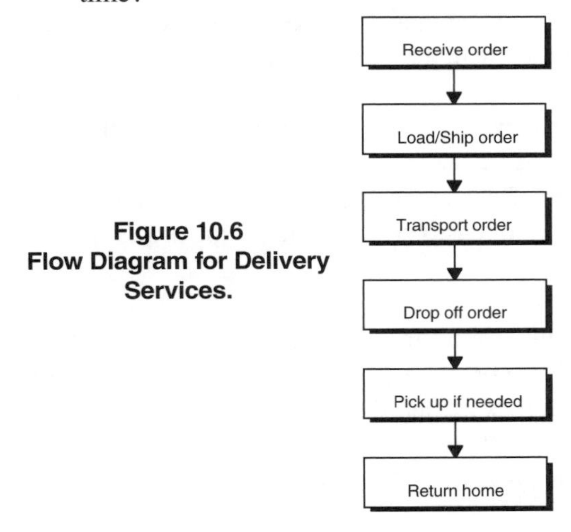

Figure 10.6
Flow Diagram for Delivery Services.

Model Representation

System element	*Model representation*
Products, loads	It is important to define attributes for products in order to keep track of unique characteristics such as weight, size, destination of the delivery, cost of the product, etc.
Queues	The merchandise or products waiting for delivery are held in queues in multiple capacity locations. Typically the FIFO queuing rule is used.
Vehicles, drivers	These are resources that are used for delivering the products to customers. These resources should be modeled as mobile resources although sometimes it may be more appropriate to model them as entities.

The roads that are traveled by these resources can be viewed as a network of paths. Distance of the paths and speed of the resources are important data items.

Arrivals

Arrivals usually result from a telephonic service. The arrival rates can be best estimated by analyzing the output rates of the ordering process.

Processes

Processes to be modeled include preparation of the product, loading of the product on the delivery resources, and unloading the product upon delivery. If there are multiple loads on a vehicle, maintaining attributes of individual entities is necessary to match the product with the right customer.

Routings

Determination of the best routing decisions for drivers may depend on the system functions such as the number of customers waiting for the product or the proximity of customers waiting for the product.

Decision logic

Programming may be needed to model complex decision rules involving scheduling of deliveries and resources. For example, if a driver has three deliveries and one pickup, and the pickup location is closest to the second delivery location, it may be desirable to pick up the load before making the third delivery.

Simulation Procedure

Simulation of deliveries is similar to simulation-based scheduling in that a simple deterministic run can be made to test a particular delivery strategy.

TRANSPORTATION SERVICES

Transportation services involve the movement of people from one place to another. A fundamental difference between transportation and delivery systems is that people are being transported rather than goods. Another important difference is that the routes in transportation services tend to be fixed whereas the routes in delivery services are somewhat flexible. Customers are interested in convenient and fast transportation. Cost of transportation plays a significant role in the selection of the service. Because set schedules and routes are used in transportation, customers expect reliable service.

Two types of pickup and drop off points are used in transportation: multiple pickup and drop off points and single pickup and drop off points. In multiple pickup and dropoff point systems, customers enter and leave the transportation vehicle independently. In single pickup and dropoff transportation, customers all enter at one place and are dropped off at the same destination.

Examples

Airlines, railroads, cruise lines, mass transit systems, and limo services

Performance Measures

- Boarding time
- Exit time
- Transportation time
- On-time arrivals
- Utilization of transportation vehicles
- Travel cost

Decision Variables

- Size and location of loading and unloading areas
- Number, size, and speed of transportation vehicles
- Scheduling of departure and arrival times
- Scheduling of vehicles and operators
- Maintenance scheduling

Questions to be Answered

- How many of each vehicle are required to meet passenger demand?
- Which transportation routes will maximize productivity?
- How many customers per vehicle?
- How can the departures be scheduled for maximizing customer convenience?
- How many shifts and service providers do we need?
- What procedures can be used (preassigning seats) to minimize service time?
- How will reliability procedures (baggage checking, etc.) affect overall service time?

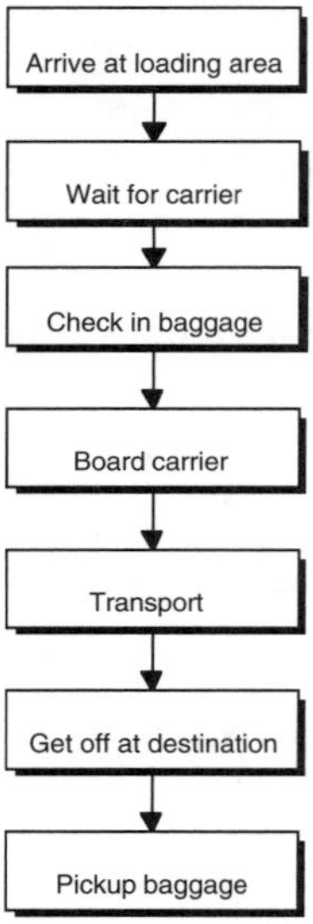

Arrive at loading area

Wait for carrier

Check in baggage

Board carrier

Transport

Get off at destination

Pickup baggage

Figure 10.7
Flow Diagram for Transportation Services

Model Representation

System element	***Model representation***
Passengers, baggage	Passengers and their baggage are modeled as entities. Customers sometimes can be classified into groups such as coach, first class, etc. It is important to define attributes for customers in order to keep track of unique characteristics such as the baggage, destination, or class type.
Queues	Customers waiting for departure are modeled as entities in a queue. Typically FIFO queuing rule is used.
Vehicles, crew	These are resources that are used for loading, serving, moving, and unloading passengers. These resources should be modeled as general use resources although sometimes it may be more appropriate to model them as entities. Modeling crew members as entities may be all right since the crew may leave the aircraft just like passengers at the termination of a trip. The roads or tracks that are traveled by these resources can be viewed as a network of paths. Distance of the paths and speed of the resources are important data items. When modeling railroads, contention for rails and collision avoidance needs to be modeled.

Arrivals	Passengers arrive at a loading area based on a predetermined departure time. The arrival rate can best be represented with an exponential distribution.
Processing	Processing of entities begins with loading of passengers and their baggage. Because there are multiple passengers in a vehicle, maintaining attributes of individual passengers is necessary to match the passenger with the right baggage at the destination. Once the passengers are all aboard, the vehicle begins movement to the destination. The vehicle may make intermediate stops to pick up/drop off passengers before it reaches its final destination. When the vehicle reaches the final destination, passengers exit the vehicle. Preparation of the resources such as vehicles (maintenance) and crew (changeover) are important processes because their delays may introduce delays for the passengers.
Operations	Boarding and exit of passengers, loading and unloading of baggage.
Decision rules	Scheduling of vehicles and crew may be based on complex algorithms. When modeling boarding activities, decision rules may be used to bump late passengers and provide room for stand-by passengers. Programming may be needed to model complex decision rules involving the scheduling of resources.

Simulation Procedure

Transportation systems are usually transient in behavior and are therefore modeled for a cycle period. Multiple replications should be made.

SUMMARY

Service systems provide a unique challenge in simulation modeling due to the human element involved. Service systems are highly labor intensive. The customer is often involved in the process and, in many cases, is the actual entity being processed. In this chapter, we discussed the aspects that should be considered when modeling service systems and presented methods for modeling different situations. Different types of service systems and modeling issues associated with each were discussed.

Chapter 11
Manufacturing and Material Handling Applications

INTRODUCTION

This chapter provides summaries of successful simulation applications within various manufacturing industries. The application summaries are presented in a common format for easy reading and quick reference purposes. Each application summary starts with a background and the purpose of the simulation study. Then, a description of the model is presented with a process flow diagram and a facility layout. At the end of each application summary, the results of the simulation project are presented. With the exception of a few, most of these simulation applications have been previously printed or presented, for which references are provided. Below is a list of the simulation application summaries presented in this chapter.

Manufacturing Industry	Type of System
Aerospace	Job shop
Automotive	Line flow
Consumer Products	Line flow
Consumer Products	Flow shop
Consumer Products	Flow shop
Electronics	Line flow
Electronics	Flow shop
Medical Device Manufacturing	Line flow
Metal Fabrication	Flow shop
Printing	Job shop/flow shop
Warehousing/Distribution	Order picking

APPLICATION 1: THROUGHPUT ANALYSIS FOR AIRCRAFT MAINTENANCE

Industry: Government (Reedy 1993)

Background: The U.S. Air Force Material Command's Sacramento Air Logistics Center (SM–ALC) was assigned with the task of Programmed Depot Maintenance (PDM) of the KC135 Air Refueling Tanker. Due to the age of the aircraft, this task required modifications to the electrical systems and extensive depot maintenance of up to 25,000 man hours. With increasing workloads that compete for many of the same resources at SM-ALC, simulation modeling and analysis were utilized. A high-level model was built to identify process and facility bottlenecks. Then, micro-level models of the bottleneck processes were built to evaluate the opportunities for improvement.

Purpose of the Study: The objective of the study was to provide a tool with which the Aircraft Directorate at SM-ALC could optimize Programmed Depot Maintenance support and provide insight regarding how best to posture the Center for major increases in workload requirements. The key performance measures were throughput, work in process, and flow days to perform maintenance on the aircraft.

Model Description: The first step in defining the model was to outline the major processes. Due to the short history of maintenance performed in the KC135 at SM-ALC and high variability in the process times, data needed for the model was collected through interviews with mechanics and work leaders. In the simulation model, triangular distributions were used for representing the process times. Figure 11.1 shows the process sequence.

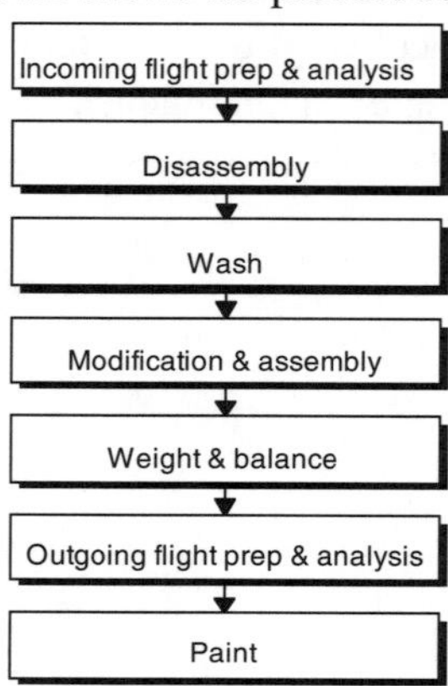

Figure 11.1 Flow Diagram of Aircraft Maintenance Operation.

When an aircraft enters the system, certain components are removed for repair or replacement. These include the landing gear, flight controls, panels, engines, and electrical wire harness. The repair times for these components occur in parallel but off the main production dock. They represent potential constraints to the smooth flow of aircraft depending on the repair requirements and material supportability. Consequently, these components were modeled as unique entities which were removed, repaired, and queued for reassembly.

A KC135 arrives every 250 hours. Once an aircraft lands, it queues for Incoming Flight Prep (IFP) and is defueled. Once defueled, the aircraft queues for entrance into one of the eight high bay docks for flight control removal and preliminary disassembly. Following this disassembly, the aircraft waits for available space in the wash/depaint facility. After the aircraft is washed, the aircraft is moved back to the high bay dock for disassembly and repair or replacement of other components such as the landing gear, engines, and panels. Then, the aircraft is moved to the refueling area and checked for leaks and flight readiness. If all systems are go, the aircraft is defueled and queued for the paint shop. At the paint shop, the KC135 competes for a turn with the F111, the F15 and the A10. Although the capacity of the paint shop is four small aircraft, when a KC135 is in the paint shop only two small aircraft can be in there. After the painting process, the aircraft queues for refueling and for delivery. At this point, the KC135 PDM is completed.

Figure 11.2 shows the facilities where the maintenance is performed.

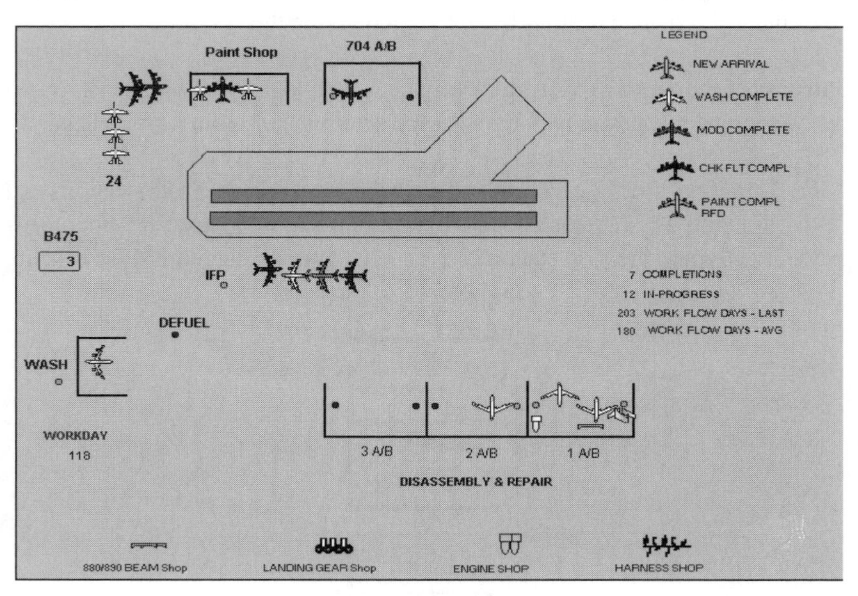

Figure 11.2 Model Layout of Aircraft Maintenance Operation.

Results. Various experiments were conducted with the model to evaluate alternatives. The experiments showed that by increasing the work week to six days, adjusting the shifts, changing the process locations for paint and reducing the work load to ten aircraft, up to 26 aircraft can be serviced with 147 average calendar flow days per aircraft. This was a significant improvement from 16 aircraft and over 200 flow days which was the current state performance. The model building, validation, experimentation and presentations included people ranging from the AFMC commander to mid-level managers to aircraft mechanics. One of the single most important benefits was the communication of the bottlenecks and the flow of aircrafts through the SM-ALC.

APPLICATION 2: DESIGN OF A BRAKE VALVE ASSEMBLY AND TEST SYSTEM

Industry: Automotive (Mabrouk 1993)

Background: A leading brake manufacturer for the automotive industry planned to purchase a new valve assembly and test system estimated to cost over five million dollars. With this capital investment, the designed line needed to be verified to ensure that it was capable of meeting production requirements. During the design phase, questions and concerns regarding the operation began surfacing among the design team. The team used simulation to evaluate alternative designs before implementing the system.

Purpose of the Study: To determine shuttle system logic, verify system throughput, determine manpower requirements and determine the number of pallets.

Model Description: The system has eighteen assembly operation stations and seven to eight test stations. A sub-assembly operation, located off-line, feeds parts directly into the load station. Figure 11.3 shows a diagram of the system.

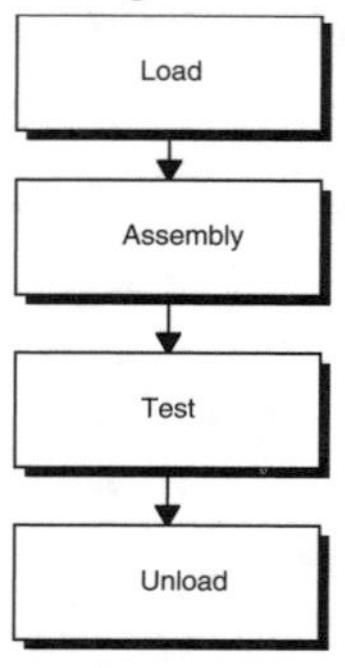

Figure 11.3 Flow Diagram of Brake Valve Assembly and Test Operation.

A gamma distribution with a shape parameter of two is used to define the time between failures (TBF) and time to repair (TTR). The mean values for these distributions were provided by the advanced manufacturing engineer. Scrap is modeled at the test stations only. For all stations on the assembly line, a down time repair cycle cannot start until both the appropriate panel mate and a maintenance person are available. For breakdowns of the conveyor or the subassembly, only a maintenance person is required. A layout of the model is shown in Figure 11.4.

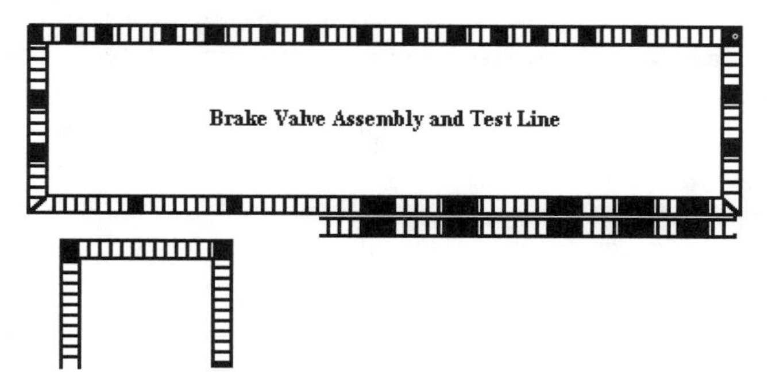

Figure 11.4 Model Layout of Brake Valve Assembly and Test Operation.

In the base scenario model, one person does all maintenance and one person performs all material handling. The maintenance person is responsible for repairing failed equipment. The material handler is responsible for loading feeders and trays, and unloading trays.

The load/unload tray system has space for 14 trays. The number of parts per tray is 24. At pallet load/unload position, a tray must move totally out of the way before another tray can start moving into the load/unload position. A tray will take four seconds each to index into and out of a load/unload position, and to index from the first to the second position in the unload queue.

If two trays happen to be queued between the load and unload positions, then these two trays must be indexed into the tray unload area (one full and one empty), once a tray is filled up at the unload position. At the unload station, test stations' rejects are dropped into a reject chute.

The capacity for most feeders is 3,600 pieces. The only exception is station two, where the capacity is 6,000 pieces. The level at which a "request to refill" is sent to the material handler(s) is 300 pieces. The time to load a feeder for a first bag in a "fill" is longer than the next four subsequent bags. This time includes the time for the material handler to get to the station, go to the racks and retrieve a bag, and empty the bag into the bowl. It also includes the time wasted when the right product is not available in the racks.

Trays are unloaded four at a time. The number of trays loaded will be as many as is needed to fill the input queue. After every 64 trays, the material handler spends five minutes moving pallets. When material handlers have no pending requests to service, they fill bowls that are less than 50 percent full.

Starting with the initial scenario, experiments were performed to determine the effect of: (1) combining the material handling and maintenance duties, (2) varying the number of operators, (3) refilling bowl feeders when they are half full (versus waiting until there are 300 parts left in the feeders), and (4) varying the number of pallets.

In the original set of scenarios, the test stations operate in one of two modes. Eight percent of the days (1 day=18 hours), the test stations reject parts at a rate of 5.9 percent. The other 20 percent of the days, the test stations reject parts at a rate of 9.9 percent. When a reject occurs, a retest is performed. Fifty percent of the parts retested are rejected. In addition, the cycle time is defined to be 50 seconds.

In early July, it was determined that the cycle time for the test stations is 55 seconds (versus 50 seconds). In addition, the first time reject rate, at all times, is estimated to be 9.3 percent. After retest, half of these first time rejects are scrapped.

At this point, as a result of the first set of experiments, it was determined that the material handling and maintenance duties should be combined, the number of operators changed to 2, and the bowl feeders refilled when they are half full. For this second base scenario, experiments were performed to determine the effect of varying the number of pallets and varying the number of test stations.

Results: The simulation results helped in arriving at a system design for producing 431 parts per hour. The model showed that this throughput could be achieved with 60 to 70 pallets in the system, 8 test stations, and 2 operators sharing both the material handling and the maintenance duties. In order to maintain this throughput, bowl feeders must be refilled as soon as they are half empty. For the startup phase of the implementation, the same setup with 4 test stations was used to produce 219 parts per hour.

APPLICATION 3: CHANGING FROM BATCH PRODUCTION TO CONTINUOUS FLOW MANUFACTURING

Industry: Metals (Marmon 1991)

Background: Teledyne Allvac is one of the world's foremost manufacturers of nickel and titanium super alloys. Teledyne produces a line of hot-rolled bar products via batch processing for its finishing operations. The company decided to migrate to continuous flow manufacturing (CFM) which would allow individual movement of products through the processes in a continuous fashion via conveyors instead of movement in batches. However, before Teledyne engineers could justify the new design, they needed to answers questions like:

- How much work in process (WIP) would be eliminated with the new system?
- How should the transfer and accumulation conveyors be configured?
- By going to CFM, how much additional production capacity will be available?
- What will be the cycle time for orders?

Purpose of the Study: The purpose of the study was to help design and justify the new inline finishing cell. It was necessary to determine the system throughput capacity and total cycle time from customer order to complete processing for a specific product mix.

Model Description: The objective of the finishing process is to bring a hot-rolled round bar into final size and surface finish requirements. The hot-rolling process produces a bar from 0.500 to 4,000 inches in diameter and up to 50 feet in length. After cooling to ambient temperature, the bars queue for the finishing process. The operations contained in the finishing process are straighten, peel, straighten, polish, inspection, and cut-to-length. This process is able to produce a bar within +/- 0.002 in. dimensional tolerance and below 63 RMS (root mean squared) surface finish.

Because of uneven cooling after the hot-rolling and heat treat processes, the bar enters the finishing operations in a bowed and bent condition. In the first finishing process, a rotary straightener is utilized to form a straight bar. The next operation is peeling. In the peeling operation, the bar is cut with carbide tooling to within final diameter specifications. This operation can remove up to 0.100 in. of material from the diameter in one pass. The straightening operation must be performed again after peeling to correct any bowing that may occur. The polish operation is used after the second straightening operation to remove all tooling marks and scratches from the bar. The polish operation assures a surface finish of at least 63 RMS, which alleviates any future cracking and allows for non-destructive defect testing.

After the polish operation is completed, the bar is inspected for internal and external defects. Once these operations are complete, the bar is cut-to-length as prescribed by the customer and then shipped. A flow diagram of the finishing process is shown in Figure 11.5.

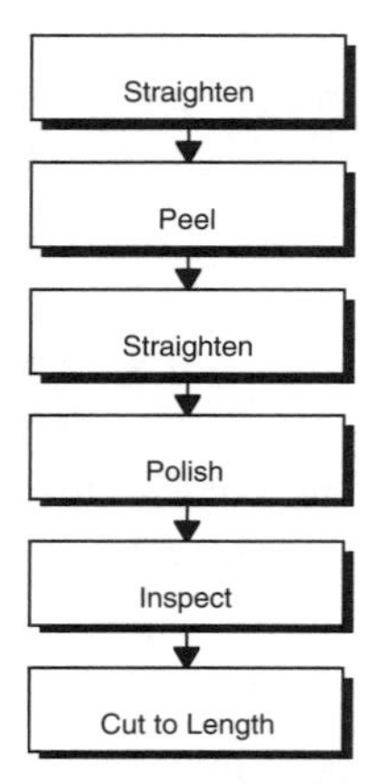

Figure 11.5 Flow Diagram of Finishing Process.

Presently, each of these operations stands alone as individual cells. An entire order is processed by each operation before it is transferred to the succeeding operation. This batch processing method is replaced in the proposed system by a flow line system in which these operations are linked together by conveyors. The use of conveyors will, in a sense, combine all of these operations into one cell. Bars will be processed individually, not as an order.

The model was built by assigning each operation only one machine. Each machine was modeled as a resource with a capacity of one. This implies each bar will be processed one at a time by each machine. After a bar finishes processing by a given operation, it will begin queuing for the next operation. The conveyors were modeled as queues with variable capacity. For this system, a large queue was considered to have a 50 plus bar capacity. A medium queue contained a maximum of 20 to 30 bars. A small queue had a capacity of 5 to 10 bars. The capacity for each queue was determined by the user based on statistics gathered from previous runs. An operation cannot release a bar into a queue if it is filled to maximum capacity. The operation must wait until queue capacity is available.

Machine cycle times and downtimes were modeled with the use of part attributes, variables, if-then logic, and built-in constructs. Machine cycle time was based on capability in feet per minute. If the cycle time was alloy-or size-dependent, a part attribute would help to determine the cycle time. Major and minor machine setups were modeled using if-then logic. Once a machine detects it has processed the last bar in an order, the machine asks for information on the next upcoming order. Based on the new order's characteristics, if-then logic lets the model determine the correct type and associated downtime for a setup. Tooling changes and machine breakdowns were modeled using the downtime constructs in the language.

One of the most important aspects of this simulation was scheduling orders into the system. Customer orders range from 500 to 25,000 pounds each. The number of orders per day can vary from 0 to 18. In other words, the system can be virtually empty or loaded to capacity. The ability to correctly model these variances was provided by user-defined as well as standard built-in functions and distributions. The model lets the user customize the scheduling module by using several distributions to enter orders into the cell. Once these orders were received, a variety of functions and distributions determined the number of bars contained in the order. Without this flexibility, real-life occurrences could not have been correctly modeled. A diagram of the model layout is shown in Figure 11.6.

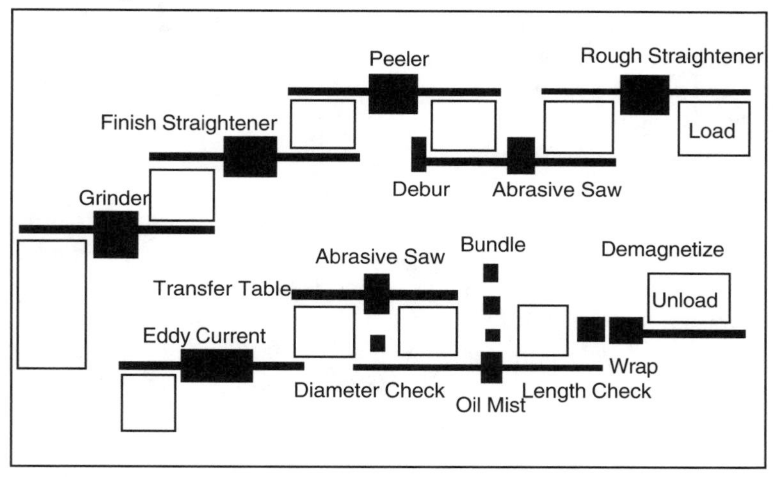

Figure 11. 6 Layout of Finishing Process.

Because of the nature of inspection processes, special consideration must be given to the queues preceding these operations. If a manufacturing defect is detected in an inspection operation, it would be advantageous to correct the problem before any more material is processed. Therefore, inspection queues should be minimized to improve information feedback to preceding operations, hence reducing defects and scrap. The queue size for each operation was optimized by analyzing statistics such as average minutes per entry, average contents, maximum contents, percent wait for output from queue, and percent wait on location. The general optimum queue sizes by operation were: straightener, small; peeler, medium; straightener, medium; polish, medium; inspection, medium; and cut-to-length, small. Output from this layout only differed by five percent compared to the layout with all large queue sizes. Once these queue sizes were determined, final statistics were generated for justification purposes.

The utilization reports show a wide range in machine utilization percentages. This behavior is consistent with today's philosophy of not running equipment if it is not necessary. Artificially inducing high machine utilization only causes an increase in WIP while output remains unchanged. The model results show a wide range of statistics for average minutes per entry and average contents. Line balancing could be used in some instances to smooth these variations. However, line balancing was not the objective because the cell is designed to run multiple products of varying diameter and alloy. It is unrealistic to attempt line balancing to reduce delays with so many product variations. Delays were minimized by customizing queue size for each operation.

Further analysis of the results indicated that the high percentage of downtime is attributed to the large number of setups required because of the product mix. Setups account for 35 percent of the total processing time for an average order. The high percentage of time waiting in queue for the first straightener and polish operations reflects the fact that the peel and inspection operations are slow in comparison.

Results: The simulation study offered two very important performance measures that could not be accurately determined using conventional methods: expect output in pounds per year and total cycle time for each order. The model demonstrated that the new design would increase capacity by 60 percent and reduce average cycle time per order by up to 90 percent. The model also allowed minimizing the queue sizes which would result in reduced floor space. In summary, simulation helped justify the investment.

APPLICATION 4: MULTIPLANT MANUFACTURING AND DISTRIBUTION SYSTEM

Industry: Consumer products (Hein and Jones 1991)

Background: A large manufacturer of consumer products was faced with the problem of satisfying increasing demands for its multiple products at multiple shipping locations. This particular company has six manufacturing and six regional distribution facilities in the United States. The customers of the company are wholesalers and retailers that ultimately distribute the products to consumers. As demand for the products increased, management needed answers to the following questions:

- Should stock be manufactured for inventory or should facilities be expanded to manufacture more products faster?
- How can delivery time and cost be minimized?

Purpose of the Study: The purpose of the model was to evaluate strategic manufacturing and distribution decisions for the multiplant environment with very large Stock Keeping Units (SKUs). The major variables of interest were cost, manufacturing capacity, warehousing capacity, and customer service levels.

Model Description: The model used a make-to-stock production system based on a forecast derived from the manufacturing planning and control system. Shipment of any product was allowed from any or all of the manufacturing facilities to any one of the multiple distribution centers or to customers. The products manufactured are all grouped in part families based on shared processes. The manufacturing process is one that required considerable reliability modeling because of the maintenance, setup and changeover requirements. It was also important to analyze the tradeoffs among capacity limitations, outsourcing and the investments in additional capacity or warehouse space.

In most manufacturing industries, such as the one described here, which involves a large numbers of SKUs, parts are grouped together into part families rather than treated individually. In order to demonstrate the need for such an approach, and the possible benefits, products were divided into the categories of continuously produced items, intermittently produced items, and critical or problem items that may be produced continuously or intermittently.

Initial assumptions were made for the model development process: the model would consider only one product at a time (a customer order would be for one product, but could include up to ten configurations or variations of that product) and some locations distribute a given product but do not produce it. Each plant is linked to a "producing" plant which ships the finished goods to its assigned "non-producers." The customer demand for the non-producers is included in the producing plant. The non-producing plant is no longer a distribution center. Figure 11.7 shows a flow diagram of the model.

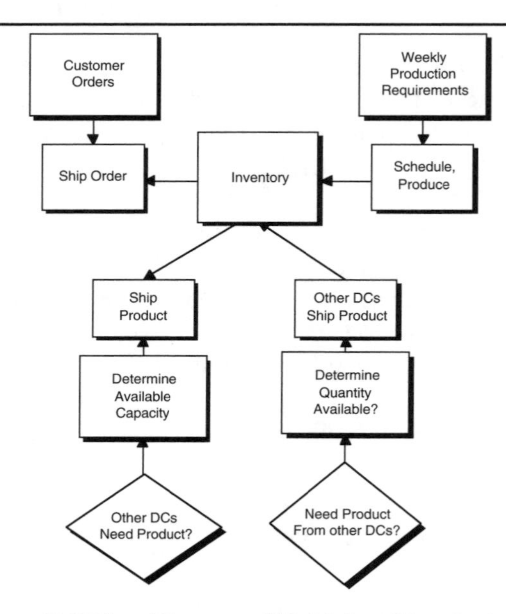

Figure 11.7 Flow Diagram of Multiplant Manufacturing and Distribution Facility.

The data requirements for this model were significant. In order to simulate three months of activity for one product (with ten variations), 2,340 data items were needed just for customer demand and target inventories. To aid in the retrieval of this data, mainframe and PC programs were written to pull this data from the MRP system and format it as required by the model. Data gathering activity required interviews with key production, financial, and shipping personnel. The following are examples of some of the information obtained:

- Cycle times
- Scrap rates
- Breakdown frequencies
- Maintenance and repair procedures
- Inventory carrying costs
- Production costs
- Shipping costs and times
- Shipping procedures
- Outsourcing procedures
- Scheduling and operating procedures
- Order and delivery cycles and procedures.

Inventory operating levels and procedures are based on the forecast of customer demand. A typical inventory operating level represents approximately one to two months of customer demand. Each producing plant has an inventory operating level that changes weekly. Each plant would like to reach but not exceed its inventory operating level. Another inventory parameter, minimum inventory, is used to determine the shipment of finished goods among distribution centers. The operating and minimum levels are used in the Manufacturing Planning and Control System (MRP system). Utilities were developed to integrate the MRP system into the simulation by downloading forecasts and inventory parameters into the simulation just as they are used to operate the manufacturing and distribution facilities.

An order history analysis was performed to build an order generator for the simulator. Typically, orders include more than one product. Since the simulation would be performed for only one product per run, it was necessary to break orders into product components, and to determine the statistical patterns for each product or develop actual order histories for each product. Several years of order histories were reviewed and a subset was analyzed in detail in order to develop a statistically based order generator for the simulator which could be used as a substitute for actually downloading orders from the MRP system. Once validated, the simulator could be an operational tool by using actual orders from the MRP system.

There are many procedures and practices that affect the rules built into the simulation model. The criteria for making decisions can change on a weekly or even daily basis. In these circumstances, decisions are made based on past experience or even an employee's intuition. Simulation models require the rules for specific situations, not intuition. Defining the specific situations and obtaining consensus on the actions to these situations was time consuming, but necessary to develop definitive procedures and operating rules. Several examples follow.

Shipments of finished goods between two manufacturing plants are made only when a producing location has a stockout (zero inventory of a particular configuration). Rules were developed to ensure that the shipping plant would not deplete needed supplies for customers while filling secondary orders for another plant. The impact of these rules becomes one of the subjects of investigation in the simulation model.

Various items are input to the model from the user interface and from the files downloaded from the MRP system. Operating rules represent predefined procedures to cover manufacturing operations, inventory policies, service level targets, priorities, and alternatives for outsourcing or backup manufacturing at substitute plants. Plant capacities and locations are defined for manufacturing and distribution. These items are used by the simulator to produce the results for a given simulation period such as a month or a quarter or a year.

Results: The model provided the company with a strategic policy-making tool to examine the impact of changes in production capacity and inventory levels on service level and cost. For example, suppose the company has a defined customer demand stream and the service level is below customer expectations. The company would make incremental runs, increasing the number of machines available or increasing target inventory levels or both, until the desired customer service level is reached.

APPLICATION 5: CELLULAR DESIGN FOR OPHTHALMIC LENS MANUFACTURING

Industry: Consumer products (Lewis 1993)

Background: SOLA is the world's largest manufacturer of plastic lenses for spectacles and sunglasses. SOLA produces many types of lenses ranging from semi-finished and finished plastic to semi-finished and finished glass. Prior to 1990, SOLA had only produced CR-39 plastic lenses in its main plant. The process was a large batch process with typical batches of 8,000 to 12,000 parts. These batches were moved on trolleys holding 1,000 lenses each between the four major processes: casting, coating, inking, and packaging. In 1990, SOLA's R&D group developed a new High Index (HI) plastic to be used for semi-finished HI lenses. Because the chemical formulation was radically different from the CR-39 material, manufacturing engineering had to develop a new approach to take this process from pilot production to high volume production.

Purpose of the Study: To determine the expected throughput of the new production process.

Model Description: The model consisted of four main processes (see Figure 11.8). The casting process was modeled as a continuous process. Molds were filled, cured, and opened. Empty molds were routed back to the filler, and the lenses were sent to the next process, which is coating. After coating, the lenses were sent to inking and then to packaging. Because the production process was completely new, the data sources for the model were the R&D chemist, the production manager, and the senior production technician. The movement between each process is done by a conveyor system moving one foot per minute. Each machine is manned by an operator.

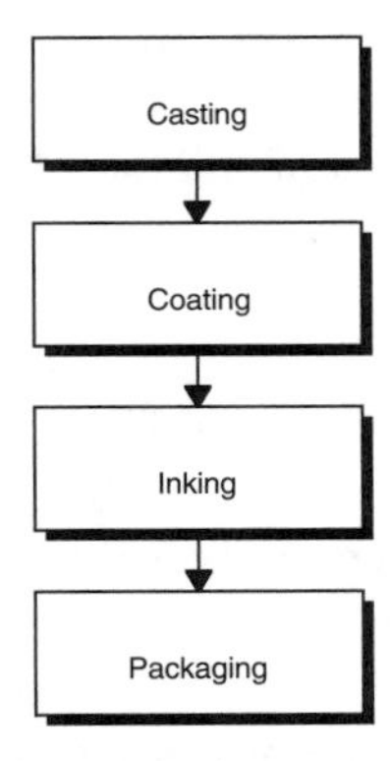

Figure 11.8 Flow Diagram for Ophthalmic Lens Manufacturing.

The modeling and analysis was an iterative process that lasted about six months. The model was initially used for buffer analysis between equipment. Initial model results showed throughput figures that were disappointing to management because they were only 40 percent of the anticipated throughput. Once actual production began, it became evident that the process could only produce 60 percent of the model's results. This proof of model validity was enough to convince management that simulation should be used to make improvements to the system. So, the manufacturing engineer began using the model with production supervisors and the operators to evaluate alternatives. Because of the past emphasis on batch processing, the operators tried to run each operation at 100 percent capacity. This caused major bottlenecks. Once the operators watched the model run, they observed the bottleneck operations. The operators in the bottleneck areas were trained to match the throughput of the model (simulating JIT). This resulted in the elimination of bottlenecks. Management accepted the value of modeling as a process definition tool, and two additional lines were built with the model data/results (see Figure 11.9).

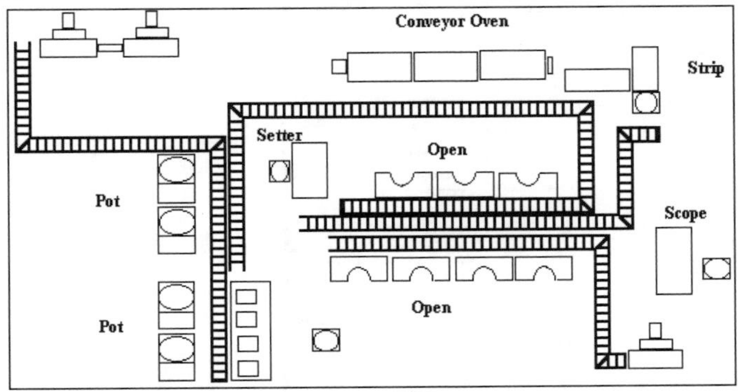

Figure 11.9 Model Layout of Ophthalmic Lens Manufacturing.

Results: The total capital investment in the line was $1.5 million. A five minute cycle time reduction results in a savings of $250,000 per year. Doubling the volume saved about $2 million which would have been the addition of a fourth line. The labor savings of not having the fourth line was estimated at $1.5 million annually.

APPLICATION 6: AN SMT PRODUCTION LINE DESIGN

Industry: Electronics (circuit board production) (Kiran, Roberts, and Strudler 1993)

Background: Hughes Network Systems was planning to design and implement a new Surface Mount Technology (SMT) line in 1990. Before automated process and material handling equipment was purchased and the facility was operational, manufacturing engineering management decided to use simulation to ensure proper design and efficient operation of the line.

Purpose of the Study: The main objective of the simulation study was to evaluate alternative systems and layout designs. The simulation study was performed in two phases. In the first phase, a basic line configuration was developed to address initial design issues and select the best line design. In the second phase, the selected line design was fine-tuned and optimized.

Model Description: The line consisted of state-of-the-art high speed component placement machines, conveyorized material handling, wave solder and cleaning machines, manual assembly, and final assembly. Figure 11.10 shows a flow diagram of the model.

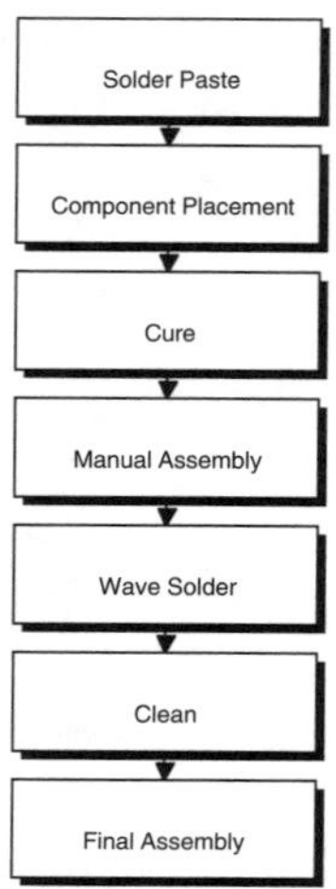

Figure 11.10 Flow Diagram of SMT Production Line Design.

Bare circuit boards enter the line at the load station and go through a component placement and infrared oven (see Figure 11.11). Some boards may have SMT components on both sides. If the secondary side of a board needs placement, boards are inverted and sent through another placement machine and infrared oven. Otherwise, they are sent to either manual assembly or final assembly.

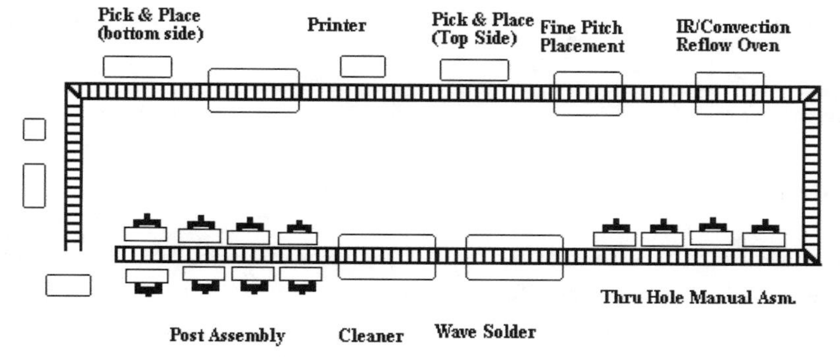

Figure 11.11 Model of SMT Production Line.

In this production system, the process equipment can be classified into two classes: component placement, paste, and epoxy application machines; and infrared ovens, wave solder and clearing machines. On the first type of equipment, process times are based on factors such as board size, number of components per board, and screen wipe. To model the process times, part attributes were used to define the functions affecting cycle time. Below are the data items represented in the model for the first type of process equipment:

- Number of components (primary side)
- Number of components (secondary side)
- Insertion rate
- Derate factor
- MTTR for placement machines
- MTBF for placement machines

The second type of machines was modeled as conveyors. The machine length and belt speed were represented in the model. The model was run for two product families. In order to study different output scenarios, the basic model was modified with various downtime assumptions (machine malfunction, maintenance, or replenishment of components).

Based on the conclusions drawn from the first phase, several additional alternative line designs were developed and tested. One of the alternatives featured an additional manual assembly station.

Results: The results from the initial phase proved that the proposed buffers by the vendors did not necessarily help balance the line. Although proposed buffers could increase the line throughput in the cases of short downtimes, they did not help keep the production rate at a maximum level in the event of extended downtimes. Subsequently, the line configuration was changed to achieve maximum throughput. This change resulted in a cost avoidance of $50,000, throughput increase of 10 percent, and cycle time reduction of 50 percent.

APPLICATION 7: EQUIPMENT DESIGN AND JUSTIFICATION

Industry: Semiconductor Manufacturing (Mauer and Schelasin 1993)

Background: It is estimated that equipment costs comprise 74 percent of the total capital costs in a semiconductor factory). Considering that a typical semiconductor tool costs $500,000 to $2 million and the average product life is only three years, maximizing capacity with minimum equipment can save millions of dollars. IBM Microelectronics, the world leader in manufacturing of semiconductors, uses simulation as the primary tool to evaluate the design and justification of new equipment. IBM was faced with a problem to meet growing production requirements in the photolithography area.

The photolithography sector is the heart of a semiconductor manufacturing line. It is also the most likely place for a bottleneck to occur, the location of the greatest variation in rework, and home to some of the most expensive tools. Knowing the capacity of this sector is tantamount to knowing the capacity of the whole line. Yet the capacity can change abruptly through rework, or be slowly modified by the simple policy of using send ahead, or lead wafers to test the performance of the optical stepper/track integrated tool.

Purpose of the Study: The options were to purchase an additional photo cluster tool, (estimated to cost $3 million) or to modify the existing equipment and optimize the process. A flexible simulation model was developed to mimic the process and play "what ifs" with various variables.

Model Description: The photo cluster consists of five processes that are serviced by a robotic handling device: prime, coat/bake, expose, post-expose bake, and develop. The tracks and photolithography equipment made by various manufacturers are linked together to form the integrated photo cluster.

Initially, a cassette of wafers arrives at a WIP station. The lot size is determined from a triangular distribution with a minimum of 15 wafers, a mode of

21 wafers, and a maximum of 25 wafers. When the stepper/track is available, the cassette is loaded at the input queue. The capacity of the WIP station is artificially set at 70 cassettes to handle the large variation of possible simulations. The cassettes arrive randomly, with a mean of 180 minutes between arrivals.

The stepper/track process is typical of a deep UV system. Each wafer from the cassette is processed sequentially. Initially, an adhesion promoter is applied, followed by an undercoat to prevent reflection during exposure. The wafers are baked, then chilled. Then the photoresist is applied, baked, and chilled. The prepared wafer is sent to the input buffer of the stepper, usually before the previous exposure is finished. Thus, the stepper itself is typically the bottleneck of the integrated system, but this need not be so. When the wafer is received in the exposure station, it is prealigned, leveled, aligned, and exposed. After exposure of the pattern into the photoresist, the wafer leaves the stepper, is baked to set the pattern, and chilled. The post-exposure bake has priority on the use of the robot. Then the wafer gets spray developed, removing the exposed resist. The fully-processed wafer is transported to the out-cassette.

After leaving the stepper/track, a few of the wafers are inspected at a microscope for gross failure of the resist process or optical exposure. Failed inspection leads to full rework of the job at this point. Time for the microscope inspection is variable and represented by a triangular distribution with a minimum of 10 minutes, a mode of 16 minutes, and a maximum of 20 minutes. During this inspection, the stepper/track is idle because a failure here is usually catastrophic. If rework is indicated, the stepper/track is taken down for repair. The repair time is variable and represented by a triangular distribution with a minimum of 5 minutes, a mode of 15 minutes, and a maximum of 60 minutes.

Following microscope inspection, the overlay of the current pattern to previous patterns is measured for each wafer; process time, 3 minutes a wafer. Then, the critical line widths are measured on each wafer; process time, 1.4 minutes a wafer. Information from these measurements is used to make a partial rework decision for some fraction of the jobs. If rework is required, a random number of wafers is selected for rework.

If rework is necessary, the cassette is sent to the plasma strip tool where the photoresist is stripped off appropriate wafers; process time, 1.2 minutes a wafer. Then residual photoresist is removed in a spray cleaner; process time, 30 minutes a cassette. The cassette with both finished and cleaned wafers is sent to a rework queue that has priority over the incoming wafers. If rework is not necessary, the cassette is taken to an out queue following the line width measurements. Figure 11.12 shows a flow diagram of the system.

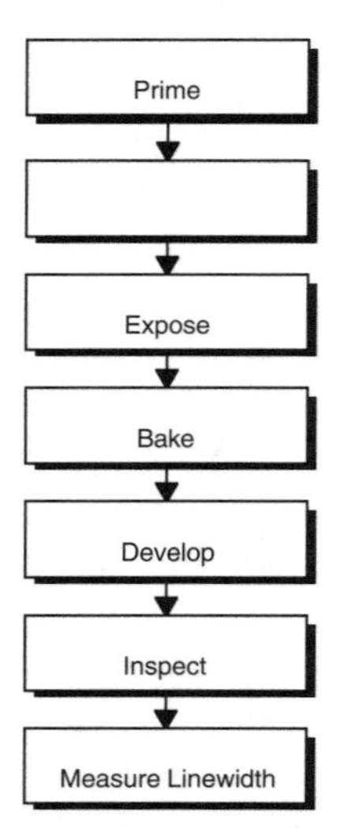

Figure 11.12 Flow Diagram of the Photolithography Process.

This process is complicated by the practice of send ahead, or lead wafers. One or two wafers are processed through the stepper/track and inspected before the rest of the lot is processed. This model uses two wafers for send aheads, allows for normal rework at both the microscope and line width measurement tools, and returns the wafers to the stepper/track. If the send ahead wafers are reworked, the send ahead procedure is repeated. If the send ahead wafers are good, the entire job is processed with the subsequent reduction of full rework to zero and partial rework to 20 percent of its previous value.

The model was developed with involvement of the tool manufacturer because data for robot speeds, internal scheduling algorithms, and tool improvements were critical to the validity of the model (see Figure 11.13).

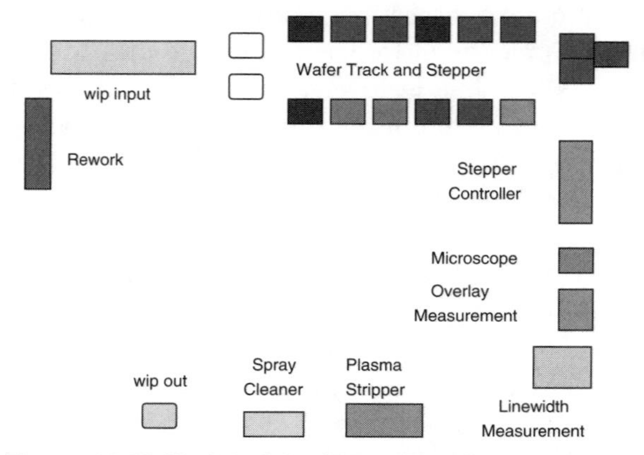

Figure 11.13 Model of the Photolithography Process.

Results: The equipment model identified several areas that could improve the photo cluster tool throughput with minimal engineering effort or qualification time. Incremental improvements were implemented the following areas:

- Increase robot handling times by varying speed and wafer transfer.
- Decrease develop cycle (prewet, develop, and rinse).
- Reduce post-expose bake chill time.
- Add a develop module (non-align levels only).

The model helped IBM Microelectronics save approximately $2.8 million (the difference between a new photo cluster versus adding an additional develop module to the existing one). The overall throughput of the existing cluster was increased by 68 percent for the non-align photo levels and 63 percent for the alignment levels. This translated into a lower tool cost of ownership and an increased level of wafer start capacity within the fabricator.

APPLICATION 8: OPTIMIZING PICKING METHODS IN A WAREHOUSE

Industry: Distribution (Guerrera and Davis 1992)

Background: The effective utilization of personnel and material handling equipment was of utmost importance to Baker and Taylor Books, a major book distribution company. Management and engineering teams wanted to study alternative methods of picking before designing the layout for a new warehouse. After realizing the importance of modeling the dynamics and randomness of the distribution warehouse, the management team assigned an industrial engineer to conduct a simulation study and provide the recommended method for picking with the optimum layout.

Purpose of the Study: The purpose of the study was to determine the optimal number of pickers and the number of totes per cart for any given picking load.

Model Description: Books are stored in three major areas of the distribution warehouse—pallet storage, flow racks, and static shelving. This bulk storage has homogenous storage locations for each stock keeping unit (SKU) as opposed to random placement of books throughout the warehouse. The most frequently picked books are placed close to the centerline path of the picker in order to minimize material movement during both the incoming storing operation and the outgoing picking operation. As the demand for particular books rises and falls, the assigned storage locations for SKUs is modified.

Picking process starts with the picker getting an empty cart from the cart staging area and walking to the pick locations on the pick list. At each picking location, the picker places books in assigned totes on the cart using the batch list. Once all the picks for the order are completed, the picker moves the cart to the unload area and returns the empty cart to the staging area (see Figure 11.14).

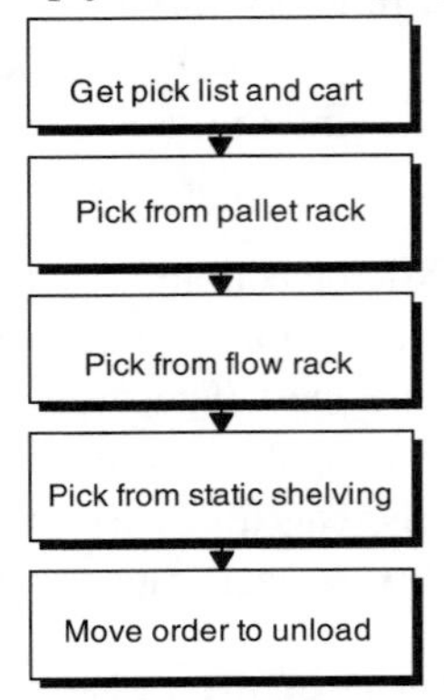

Figure 11.14 Flow Diagram for Warehouse Picking Operation.

The most critical parameter is the capacity of the picking cart which is determined by the number of totes per cart. This number affects the walking pace of the picker because the heavier the cart, the slower the pace. The capacity of the picking cart also affects the number of lines picked per trip because the larger the capacity, the fewer the trips per pick. The model starts with origination of book orders to be picked from the three storage areas. These arrivals are created according to a uniform distribution, with a minimum of 300 and maximum of 360 lines picked per trip. The model generates a random number for the total lines per trip, to determine the exact quantity of lines per trip for each storage area using probabilities defined by the user. In this case, it was assumed that 80 percent would be from static shelving, 19 percent from flow racks and 1 percent from the pallet racks. Figure 11.15 is a flow diagram of the model.

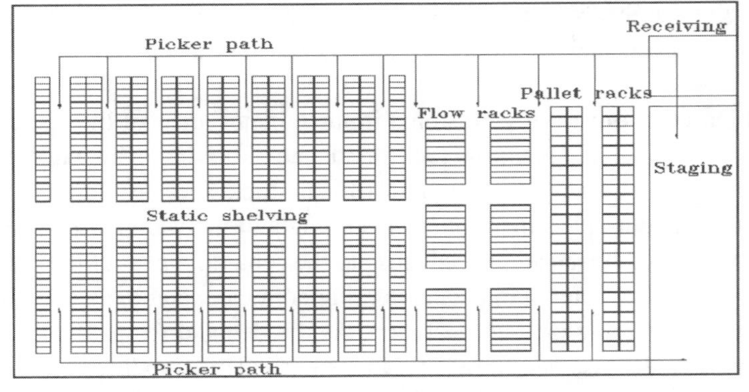

Figure 11.15 Model of Warehouse Picking Operation.

Results: The results of the model provided valuable statistics in determining the optimal number of pickers and quantity of totes per cart based on different demand. A decision matrix was developed to optimize the throughput while maximizing picker productivity. This meant that for the existing levels of demand to be met, 3 pickers picking carts with 16 totes and 3 pickers picking with 20 totes per cart would provide the most cost effective method to run the warehouse. If the demand level increased or decreased, the simulation model would be run with alternative pickers and quantities per cart to see if the demand could be met. The model was also helpful in demonstrating to the pickers the reason why they sometimes needed to pick heavier loads (20 totes per cart) instead of lighter loads (16 totes per cart) to meet the demand.

APPLICATION 9: DESIGN OF A CONTINUOUS PROCESS FACILITY

Industry: Beverage (Harrell 1993)

Background: A major beverage manufacturer was evaluating the design of a modern beverage processing facility that would enable the beverage to flow through a flexible pipe network rather than through dedicated pipe connections. The sophisticated operation and complex networking posed a real design and management challenge to take full advantage of the available technology. This type of a processing facility used to be a "make to stock" operation. The new design would enable the manufacturer to operate in a "make to order" mode. The key piece of equipment in the system is the filler (which costs about $400,000).

Purpose of the Study: The objective was to find the optimum batch size and production sequence that minimizes beverage production time. In order to achieve the objective, the following questions needed to be answered:

- Which flavor to process next and how much (batch size)?
- When to shut off the flow of input and output from tanks?
- What the rate of flow is between tanks?
- Which tank(s) should be selected for input flow?
- Which tank(s) should be selected for output flow?

The performance measures were time to process a particular mix of products, tank utilization, and throughput capacity.

Model Description: In beverage processing, beverage is processed and packaged according to some schedule or production plan. Each beverage product usually begins with either a partial or complete mixing of the ingredients comprising the beverage (sugar, flavoring, water, etc.). Depending on the batch size, this mixing may require more than one mixing tank or perhaps multiple mixes in the same tank. A production schedule was defined for 12 different beverages and 10 different container sizes, making 120 different product types.

Flavor batches were processed so as to keep as many fillers busy as possible, thus minimizing production time. The size of a flavor batch had to be large enough to meet production requirements for a particular flavor while at the same time attempting to keep all of the fillers busy. In order to exactly meet required production quantities, the batch size for a particular flavor often required only partially filling a tank at the end.

As each batch of a particular flavor is mixed, the beverage is often routed or pumped through one or more pipes and tanks until it reaches the fillers where the final containers (bottles or cans) are filled. As the beverage flows through the system, other ingredients may be blended in, such as water or carbonation. The routing sequence for a particular beverage is often dependent on which tanks, pipes, and fillers happen to be available at the time. In fact, splitting or branching the flow to alternate or even to multiple tanks and fillers is not uncommon.

After mixing beverage ingredients in an initial mixing tank, the beverage was pumped to one of six holding tanks in a round-robin fashion. The holding tanks each had a capacity of 5,000 gallons and generally required more than one tank to hold a particular flavor batch.

From a holding tank, the beverage was sent as needed for filling to one of four blenders where additional ingredients were blended in. The rule for blender selection was to give first preference to a blender that had currently been processing the same flavor, then to the first one available.

From the blender, the beverage was routed to one of ten different fillers, with each blender feeding up to two different fillers at a time that might be requiring that particular flavor. When a flavor was finished at a particular filler, other fillers of that flavor were examined for possible routing. The fill rate at each filler changed with each change in container size that was being filled.

Cleaning and changeover time was required for each tank and pipe when a new flavor was introduced. Additionally, a setup was required for fillers when the container size changed. Figure 11.16 shows a flow diagram of beverage processing. Figure 11.17 shows a model layout of a beverage operation.

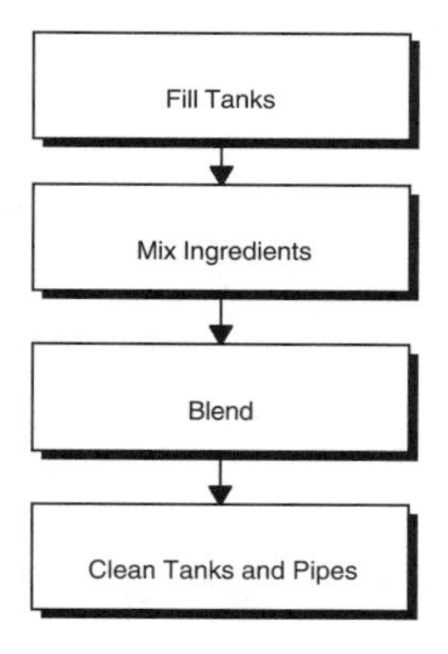

Figure 11.16 Flow Diagram of Beverage Processing.

In conducting simulation experiments, the following variables were changed to analyze their impact on the above performance measures: batch sizes, sequencing of batches, overlapping of batches, routing rules, blending amounts (when should dilution be performed and how much dilution should occur at each stage).

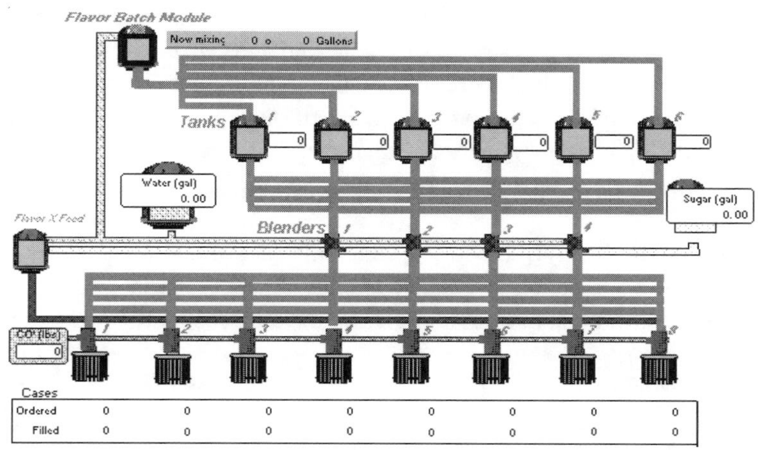

Figure 11.17 Model Layout of Beverage Production Operation.

Results. The results show that it took 1,396 hours to process the orders. The six tanks were nearly fully utilized (there was at least some beverage contents in them) for the entire simulation time. Only two of the four blenders were being almost fully utilized while the other two were hardly used at all. Products were at every filler except three for nearly the entire time, either being filled or waiting to be filled. The other three fillers were limited to special products and therefore were only used as needed.

What these results indicated, and what was especially evident from watching the animation, is that there were multiple flavors waiting to be filled, but the tanks were providing no more than two flavors the vast majority of the time. This is why only two of the blenders were being used. Another statistic of interest was the demand for water over time. The supply of water was assumed to be inexhaustible in order to find out the demand for water over time. This information was helpful in planning and managing the water supply for the system.

APPLICATION 10: MINIMUM COST TEST PLANNING

Industry: Medical Instruments (Henderson and Kiran 1992)

Background: IMED Corporation is one of the largest manufacturers of medical instruments. IMED's infusion pumps are used in hospitals worldwide. Historically, the test points (when and where the pumps are tested in the assembly operations) were determined by the production and test engineering management. The general approach in the past had been to test at the completion of each subassembly level. The failure rate at a particular subassembly level had to be almost nonexistent to cease testing at that point. The rationale for this plan was that it is less expensive to find and repair a defect at an early stage of production rather than allow the defect to progress to final test, where the test performed does not provide as much diagnosis information, and the repair is therefore longer and more expensive. Thus, the focus had been on the cost of repair and not on the total cost of test and repair combined.

Purpose of the Study: Due to the variability of test and repair data, IMED decided to use simulation to evaluate alternative test scenarios and to develop an acceptable solution that would minimize the total cost of test and repair.

Model Description: The product consists of an enclosure containing a mechanical pumping mechanism powered by a stepper motor, several circuit boards containing digital and analog circuitry, a circuit board with alphanumeric displays attached, a membrane pushbutton control keypad, a hinged door to contain the disposable pumping chamber, and clamp on the back of the enclosure to mount the pump to an IV pole or bed.

The facility consists of several subassembly areas and a final test area. All areas with the exception of final QC test are staffed by production personnel that perform both the assembly and tests according to the process specifications. The final QC test is the ultimate determinate of whether a pump meets all performance criteria and is ready for shipment. This test is performed on 100 percent of the units shipped.

The tests performed in the subassembly processes are either conducted by the assembler or a member of the statistical process control team. These tests are aimed at two objectives: to identify defects earlier in the assembly process at a point where the correction of the defect is faster to diagnose and less expensive to repair; and to identify defects that are of a nature that would be invisible at final QC test due to the unit being fully assembled at that point. Figure 11.18 shows a flow diagram of the model.

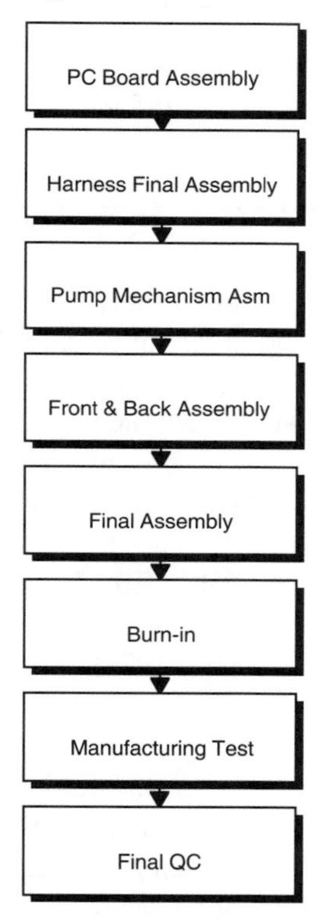

Figure 11.18 Flow Diagram of Test of Pump Assembly and Test.

Discovering defects of the first type could be deferred to later tests without compromising the ultimate quality of the product. The penalty resulting from deferral of the defect is increased diagnosis time at final QC. Defects of the second type must be found and fixed before the product is submitted to final QC.

The base model was designed and validated with historical data for test and repair times and run rates. The four performance measures of the model were average test time per unit, average repair time per unit, average WIP, and average throughput time per unit. Once the baseline model was validated, alternative test plans were simulated.

There were a total of three test plans that were simulated. The first alternative combined final assembly and burn in processes. The second alternative combined front, rear and final assembly processes. And, the third alternative combined all prior tests performed in subassembly processes into final assembly. The table below show the results of the simulation experiments.

Table 11.1 Results of Simulation

	Baseline	Scenario #1	Scenario #2	Scenario #3
Avg. test time	109 min.	106 min.	102 min.	59 min.
Avg. rework time	19 min.	19.7 min.	20.2 min.	32.2 min.
Avg. WIP	$496K	$516K	$499K	$747K
Avg. throughput time	54 hrs.	52.6 hrs.	53.6 hrs.	61.2 hrs.

Results: The simulation of alternative test plans proved that the current approach of testing after every subassembly level is the most expensive alternative. The greatest potential savings can be gained by substantially reducing the number of tests in the subassembly levels and accepting a greater amount of complex diagnosis and rework later. Therefore, the recommendation was to combine subassembly level tests into final assembly and to increase the effectiveness of the technicians with high technology test equipment.

APPLICATION 11: BUSINESS PROCESS RE-ENGINEERING PROJECT

Industry: Publishing/printing (Nedza 1994)

Background: CCH Incorporated, a leader in the publishing industry, is one of the few publishers that owns its own printing facilities. CCH owns and operates plants in three major U.S. cities: Chicago, Illinois; Clark, New Jersey; and St. Petersburg, Florida, along with other plants around the world. The printing operations require significant capital investments in equipment. For example, a new press costs about $1.5 million; a bindery line costs about $1 million.

Due to increasing pressures to maintain profitability while maximizing utilization of current assets and minimizing turnaround time to customers, CCH embarked on a re-engineering project for its plants in the U.S. The Operations Group in the Chicago headquarters was certain that the plants had excess capacity. However, the group had difficulty in quantifying and demonstrating it to upper management during the re-engineering project. With the aid of a major consulting firm, the Operations Group of CCH used simulation to evaluate alternatives and recommend solutions to upper management.

Purpose of the Study: The purpose of the simulation study was to evaluate the optimal plant size for various demand levels and to evaluate tradeoffs associated with printing various product types using current flows. One of the specific objectives was to maximize press utilization while minimizing cycle time.

Model Description: The plant operations consisted of four major operations, namely prepress, pressroom, bindery, and mailroom. For the purpose of the study, the products were classified into four major categories: current mail; reprints; small-run, low page-count books; and revisions. The difference between the reprints and revisions was the quantity and frequency of printing. The scope of the simulation study excluded operations such as compiling, warehousing and shipping. Some of the products (such as newsletters; multi-color work; large-run, high page-count books; and guide cards) were not included in the model. Figure 11.19 shows a flow diagram of the model.

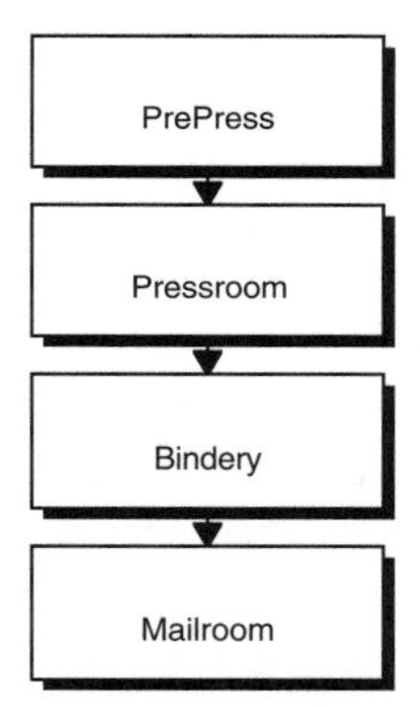

Figure 11.19. Flow Diagram of Printing Operation.

One of the unique aspects of the model was the Excel spreadsheet that was used to drive the production schedules in the model. The spreadsheet contained the following information for customer orders:

- Number of forms for each job
- Scheduled press run length for the job
- Priority for the job based on product type
- Bindery routing label based on job type
- Time that the job arrives in the job queue
- Sequential job number

This spreadsheet listed the job data for each day. Jobs arrive in the model from the spreadsheet every 24 hours. When all the forms for a particular job are ready, they are sent to the Rachwals. The Rachwals are used for creating plates for each job. The plates wait in a plate queue which acts as the buffer between prepress and the pressroom. In the pressroom, the appropriate number of forms for each job are printed. From the presses, the forms are sent to the bindery department. In the model, a few logical locations were defined between the pressroom and bindery department. One of the logical locations was the signature queue that holds printed signatures until they can be matched up with the other completed signatures for that job. The other logical locations were the queues between the initial job queue and the prebind queue. These locations were used to assemble the jobs (containing all appropriate forms). The queues contained detailed logic statements to distinguish between forms and signatures and to avoid potential blockage in the model.

The prebind queue is used for holding assembled jobs and passes them onto the appropriate bindery equipment. Of the bindery equipment, the sidestitcher stitches current mail, reprints and revisions into blocks of 12 signatures each. The perfect binder binds perfect-bound books and the saddlestitcher binds saddlestitched books. The mail from side stitchers goes to the inserters and then to the customers. The revisions or reprints from the side stitchers go to the block queue where they wait for processing at the blockgatherer. The block-gatherer gathers signatures if more than one signature is for a job.

The model included 35 variables that were defined for flexibility in the model. These variables were used for defining cycle times, setup times for presses, and number of impressions per roll of paper. The model included scheduled maintenance and random downtimes for all the equipment. The model assumed that the plants operated three shifts a day, eight hours per shift, five days a week. Paper waste was tracked by a variable called "waste" that was incremented every time a plate is changed (200 waste impressions) or every time a roll is changed (1,500 waste impressions).

Results: In summary, the study was instrumental in moving to a focused plant on core products. The model results helped CCH move to an operations mode that meets its cycle time goals with one plant with three presses instead of three plants and eight presses. The model demonstrated that peak demands on the operations caused by books and revisions can be smoothed out by outsourcing them. The sensitivity analysis with the model showed that sixty-six percent press utilizations would be acceptable to maintain the cycle time goals. Any additional utilization for the presses did not significantly affect the total throughout time.

Chapter 12
Service Applications

INTRODUCTION

This chapter provides summaries of successful simulation applications within various service industries. The application summaries are presented in a standard format for easy reading and quick reference purposes. Each application summary begins with a brief background and the purpose of the simulation study. A description of the model is presented with a process flow diagram. The results are found at the end of each application summary. A reference is provided for those applications that have been previously published or printed. Below is a list of the service applications reviewed in this chapter.

Applications	Type of System
Healthcare Services	Service Shop
Healthcare Services	Service Shop
Financial Services	Mass Service
Financial Services	Help Desk
Restaurant	Service Factory

APPLICATION 1: DENTAL CLINIC

Industry: Healthcare Services (Whitworth 1993)

Background: The Scott & White organization is located in Temple, Texas and consists of a 354 bed hospital; a 100,000 member health maintenance organization; a 375 physician, dentist, scientist, multi-specialty clinic; and a skilled nursing facility. Scott & White operates 13 regional clinics throughout central Texas. For many healthcare providers, the shift from inpatient to outpatient services has created a great many problems ranging from patient flow to space availability, scheduling, and staffing. With these types of problems, coupled with growing patient demand, the dental clinic turned to the hospital management engineering staff for solutions. After an evaluation of queuing analysis, linear programming, and simulation, the management engineers selected simulation as the best analysis tool because it allowed accurate representations of variability in daily patient demand as well as in service times.

Purpose of the Study: The objective of the simulation study was to determine an effective and efficient way to schedule patients, rooms, and dentist/oral surgeons while maximizing the throughput of the clinic.

Model Description: The dental clinic is staffed with two full-time oral surgeons, one full-time general dentist, one part-time oral surgeon, two part-time dentists, two full-time hygienists, three full-time LVNS, two full-time oral surgeon assistants, and three full-time certified dental assistants. Each dentist/oral surgeon has his or her own assigned examination room; some have one and others have two. There is no recovery room; therefore, patients are required to recover in the physician's exam room.

The data collection phase involved interviews with all personnel and analyzing the patient registration system, policy and procedure guide, and room assignment codes. In addition, a two week, on-site data collection exercise was conducted. Data was collected on:

- Dentist/oral surgeon schedules
- Patient mix
- Procedure service times by dentist/oral surgeons
- Procedure service times by patient mix
- Patient arrival rates by dentist/oral surgeon
- Patient volume by day of week
- Procedure requirement (staff type, LVN, RN, NA)
- Room assignments

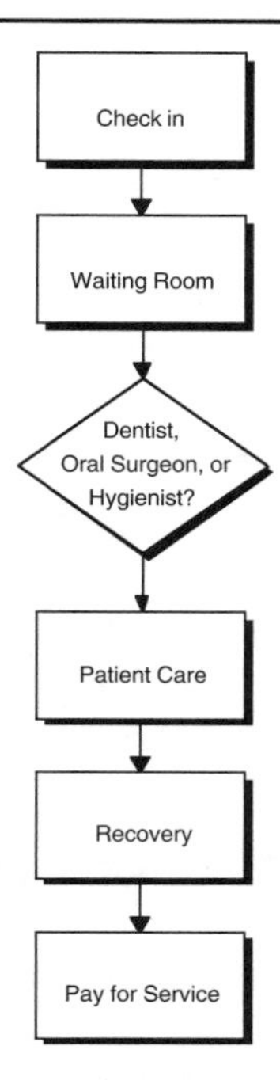

Figure 12.1 Patient Flow Diagram for Dental Clinic.

Before the model was coded, a simple flow diagram was created (see Figure 12.1). The flow diagram mapped the processes, aiding in building the simulation model. Histograms and pie charts were used to identify the empirical distributions to be used in the model. The patient mix was classified into eight types, six of which required dentist/oral surgeons and two requiring hygienists. After analysis of the data, arrival rates were represented with uniform distributions while normal distributions were used to define service times.

Because service times were selected from probability distributions based on patient type, the most critical aspect was the assignment of attributes to patients as they arrive at the clinic. Some assumptions in the model were:

- Each patient is always escorted in and out by either the LVN or oral surgeon assistant.
- No shows were not modeled.
- All patients in the clinic are provided service before the day ends.
- Patients are serviced in FIFO order when their dentist/oral surgeon is available.
- An assistant or LVN is always present with the dentist/oral surgeon when a patient is in a room.
- Pediatric dentistry was not included in the study.

Results: The analysis of the simulation results provided a number of suggestions for improving throughput. The most important conclusion was that room utilization was unacceptably low. This low utilization was found to be caused by the fact that rooms were too specialized and there was no designated recovery area for patients with IV procedures. Another contributing factor to the low room utilization was the fact that each examination room was dedicated to a particular dentist or oral surgeon.

The following recommendations were made:

- Schedule rooms by availability of rooms not by dentist/oral surgeon
- Examine the feasibility of purchasing equipment carts so that all rooms can be used by general dentists/oral surgeons.
- Schedule patients with IV procedures on those days of the week when patient volume is lower.

APPLICATION 2: EXPANSION OF AN EMERGENCY DEPARTMENT

Industry: Healthcare Services (Kraitsik and Bossmeyer 1992)

Background: The hospital at the University of Louisville serves a large indigent population and is the major teaching hospital in the Louisville area. The emergency department (ED) was originally designed with 16 treatment bays and 2 trauma rooms. The average number of patient visits per day currently ranges between 110 and 120. The current patient census and acuity level mix has exceeded the original design projections. Patient turnaround times are longer than desired. Since 15 to 25 percent of patients seen in the ED become in-patients, it is the "front door" and first impression of the hospital's quality of care for many patients and their families. This has created the need for an expansion of the ED.

When ED expansion plans were initiated, it was recognized that expansion of the ED alone would not guarantee improved turnaround times. Operational considerations needed to be included in the redesign which would ensure improved patient throughput. Simulation was chosen as the tool to examine the most cost effective strategies to reduce patient wait times and increase throughput because of simulation's ability to take into account both the probabilistic and operational nature of the problem.

Purpose of the Study: The primary question to be answered by this study was: Which of the proposed operational alternatives would best achieve the greatest improvement in patient turnaround time? Other questions that needed to be answered were:

- What are the current bottlenecks in the emergency department?
- How many treatment areas are needed?
- What would be the effect of different patient flows for hold (ICU), detox, and psychiatric patients?
- What would be the effect of a fast track?
- What is the effect of various doctor and nurse work schedules?
- What various service times were improved with increasing patient volumes?

Model Description: The patient flow is shown in the flow diagram in Figure 12.2. Specific data on patients used in the simulation was based on seven consecutive days of data collected. This one-week sample was adequate to reliably represent ED operations.

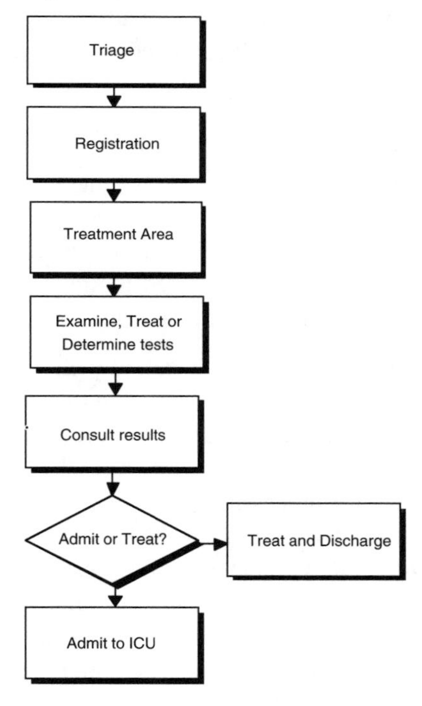

Figure 12.2 Patient Flow Diagram for Emergency Department.

The data was summarized using Lotus 1-2-3 with patient arrival rates being defined by hour of the day. The emergency department staff was consulted frequently to validate the data. Triangular distributions were used to represent treatment times because of insufficient time to gather the data necessary to warrant statistical curve fitting.

Validation of the model was performed during every phase. Model output was repeatedly compared to actual data to identify inconsistencies between the model and the actual systems.

- Paperwork for higher acuity patients takes precedence over lower acuity patients.
- Lower acuity patients will not go to the back treatment bays unless fewer than 16 beds are in use.
- No one waits longer than six hours for lab results or treatment.
- Time was included in the model for telephone calls, meals, breaks, and other miscellaneous interruptions.

The analysis of the base case (the existing situation) indicated that the availability and number of doctors were the primary bottleneck, and that this, in combination with waiting for laboratory results, accounted for the major portion of patient turnaround time. Experiments were conducted to address the impact of this and other issues.

Each experimental simulation was run for 120 days. Comparisons by acuity level were made of the time patients had to wait for a bed, the time in the treatment bay, and the "Left on Own Accord" (LOA) rate for each scenario.

Results: The results of testing several alternatives indicated that the changes to the configuration of the ED that provided the greatest potential improvement in patient throughput and quality of care are:

- Introduction of a 12-hour fast track staffed with a nurse practitioner and technician. This change resulted in reducing the length of visits for non-urgent patients to just one hour from the previous visit length of four hours.
- Use of a "stat" lab for processing high volume tests for the ED, operating room, and intensive care unit areas.
- Provision of 24 hour/day direct emergency psychiatric care, rather than 40 hours/week care. This will provide an improvement in timeliness, quality, and cost of care for patients as well as an improved ED image without the additional stresses associated with psychiatric patients.
- Increased number of treatment bays.

In March of 1994, Columbia/HCA announced a $7.5 million addition to the ED at the University of Louisville Hospital, adding much needed space to one of the premier trauma centers in the United States.

APPLICATION 3: CAPACITY ANALYSIS OF THE LETTERSHOP OPERATIONS IN A BANK

Industry: Financial Services (Nixon 1994)

Background: The Canadian Imperial Bank of Commerce (CIBC) is Canada's second largest full-service bank. CIBC has headquarters in Toronto, Ontario and a network of over 1,400 branches across Canada. CIBC offers a complete range of financial products such as personal checking and savings accounts, investment and brokerage services, commercial payroll services, group home and car insurance, credit card services, and trust services. CIBC measures its performance using a number of key factors: customer satisfaction, people management, service quality, risk management, operational capability, and shareholder value. The Operational Capability Research (OCR) group investigates, evaluates, adopts, and implements tools and techniques that enhance the operational capability throughout CIBC. The back office of CIBC, like most financial institutions, is experiencing rapid and extensive change as a result of increasing pressures to manage costs.

The OCR group selected the lettershop as the first application of simulation within CIBC. One of the primary reasons for this was because the lettershop management had already collected a great deal of process and performance data. The lettershop is responsible for processing and issuing bank statements to account holders.

Purpose of the Study: The purpose of the simulation model was to evaluate the production capacity and production scheduling options in the lettershop. The model was developed to allow the lettershop management to experiment with mail volume variability that might result from new business, and increased capacity that might result from new machine acquisition.

Model Description: Batches of customer statements for a number of the bank's products (VISA) are delivered by the processing centers twice a day. The batches are checked, sorted, and then allocated to a number of insertion machines. Individual statements are separated out of the batches combined with marketing brochures, informational mailers, return payment envelopes, and inserted into envelopes. The completed mail is then batched through an encoding machine that sorts the mail according to the post office requirements. Containers of the sorted envelopes are then shipped to the post office. A flow diagram of the statement processing procedure is shown in Figure 12.3. In 1993, the lettershop processed over 35 million items of mail for various product groups within CIBC.

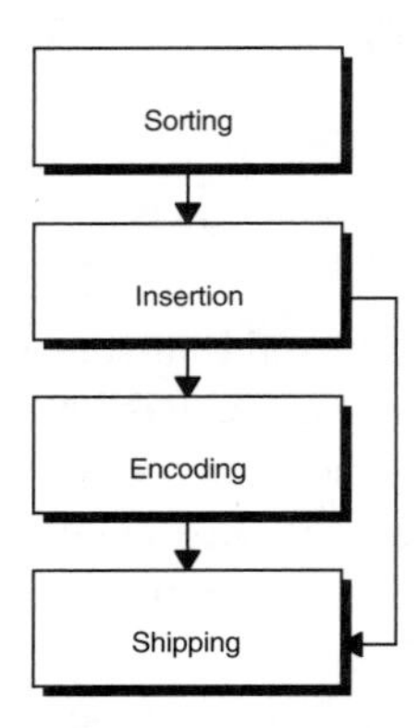

Figure 12.3 Bank Statement Flow Diagram.

The main performance measures at the lettershop are its service level agreements with its clients. These agreements feature a 24-hour turnaround cycle: statements delivered to lettershop for processing must be at the post office 24 hours later. In order to meet these commitments, the lettershop management closely tracks the number of statements processed per hour. The following data is collected on a regular basis:

- Statement volumes received (by product)
- Statements processed (by product)
- Statements left in the system
- Insertion machine utilization and downtime
- Machine setup times
- Production rates (items per hour)

In creating the model, certain assumptions were made:

- The decision-maker can select one of the 19 monthly bill days when experimenting with statement volumes or machine usage.
- The system is empty at the beginning of that day.
- The model is a macro model of the processes within the lettershop.
- The transport dollies used for moving the mail are unconstrained.
- The smallest entity or unit of processing is a *web*. A web contains a variable quantity of statements. This quantity is modeled as an entity attribute in the model. On some days, lettershop receives more than 125,000 individual statements. The model accounts for all the statements through attributes and variables.
- The higher volume encoder was modeled as a simple in-out producer.
- All insertion machines "jam" according to the same statistical distributions.

The data that drive the model came from existing production data, on-site time studies by the work measurement group and client-approved estimates. The model run length was 24 hours with the user choosing which of the 19 days the simulation should be run. The model validation was accomplished with three walk-throughs. The model flow and logic were verified using the animation features of the simulation software.

Experiments were conducted on the following decision variables to assess sensitivity and to provide answers to client questions:

- Number of insertion machines
- Insertion machine throughput capacity
- Volumes of statements received on a given bill day
- Distribution of statement arrivals (single batches versus multiple batches)
- Scheduling of machine availability (one shift, two shifts, 24 hour operation)

The lettershop modeling project took about three and a half months to finish. The project team consisted of four members who devoted anywhere from 20 percent to full time to the project.

Results: The project team learned extremely valuable lessons from this first experience with modeling at CIBC. First, the team underestimated the amount of time needed to understand, verify, and correct the production data. Second, the OCR team learned that the project specification phase must include a model specification and assumptions document that confirms a mutual understanding of the model detail and expectations between the modeling team and the client. Finally, the model provided a risk-free, robust and valid experimental lab in which lettershop management could test its decisions about capital acquisitions of new equipment and scheduling of resources for additional business.

APPLICATION 4: CLIENT SERVICES HELP DESK

Industry: Financial Services (Singer and Gasparatos 1994)

Background: Society Bank's Information Technology and Operations (ITO) group offers a help desk service to customers of the ITO function. This service is offered to both internal and external customers, handling over 12,000 calls per month. The client services help desk provides technical support and information on a variety of technical topics including resetting passwords, ordering PC equipment, requesting phone installation, ordering extra copies of internal reports, and reporting mainframe and network problems.

The old client services help desk process consisted of a mainframe help desk, a phone/local area network help desk, and a PC help desk. Each of the three operated separately with separate phone numbers, operators and facilities. All calls were received by their respective help desk operators who manually logged all information about the problem and the customer, and then proceeded to pass the problem on to an authority group or expert for resolution. The help desk acts as the primary source of communication between ITO and its customers and works with authority groups within ITO to provide work and support when requested by a customer.

Because of acquisitions, the increased use of information technologies and the passing of time, Society's help desk process had become fragmented and layered with bureaucracy, and was a good candidate for a process redesign. It was determined that the current operation did not have a set of clearly defined goals, other than to provide a help desk service. The organizational boundaries of the current process were often obscured by the fact that much of the different help desks' work overlapped and was consistently being handed-off. There were no process performance measures in place in the old process, only measures of call volume. A proposal was made to redesign the help desk functions into a consolidated help desk. The proposal also called for the introduction of automation to enhance the speed and accuracy of the services.

Model Description: The situation suggested a process redesign to be supported by simulation of the business process to select and validate different operational alternatives.

Detailed historical information was gathered on call frequencies, call arrival patterns, and length of calls to the help desk. This information was obtained from the help desk's database, ASIM. Table 12.1 summarizes the call breakdown by call level. Level 1 calls are resolved immediately by the help desk, Level 1A calls are resolved later by the help desk and Level 2 calls are handed off to an authority group for resolution.

Table 12.1 Types Of Calls For Help Desk.

Time Period	Password Reset	Device Reset	Inquiries	% Level 1	% Level 1A	% Level 2
7am - 11am	11.7%	25.7%	8.2%	45.6%	4.6%	47.3%
11am - 2pm	8.8%	29.0%	10.9%	48.7%	3.6%	44.3%
2pm - 5pm	7.7%	27.8%	11.1%	46.6%	4.4%	45.8%
5pm - 8pm	8.6%	36.5%	17.8%	62.9%	3.7%	32.2%
Average	9.9%	27.5%	9.9%	47.3%	4.3%	48.4%

Historically, calls averaged 2.5 minutes; lasting anywhere from 30 seconds to 25 minutes. Periodically, follow-up work is required after calls which range from 1 to 10 minutes. Overall, the help desk service abandonment rate was four to 12 percent (as measured by the percentage of calls abandoned), depending on staffing levels.

The help desk process was broken down into its individual work steps and owners of each work step were identified. Then, a flowchart that described the process was developed (see Figure 12.4). From the flowchart, a computer simulation model was developed of the old operation and was validated by comparing actual performance of the help desk with that of the simulation's output. During the ten-day test period, the simulation model produced results consistent with that of the actual performance. The user of the model was able to define such model parameters as daily call volume and staffing levels through the use of the model's interact box, which provided sensitivity analysis.

Joint Requirements Planning (JRP) sessions then allowed the project team to collect information about likes, dislikes, needs and improvement suggestions from users, customers, and executives. This information led to a clarification of the target goals of the process along with its operational scope. Suggestions were collected and prioritized from the JRP sessions for improving the help desk process. Internal benchmarking was also performed using Society's customer service help desk as reference for performance and operational ideas.

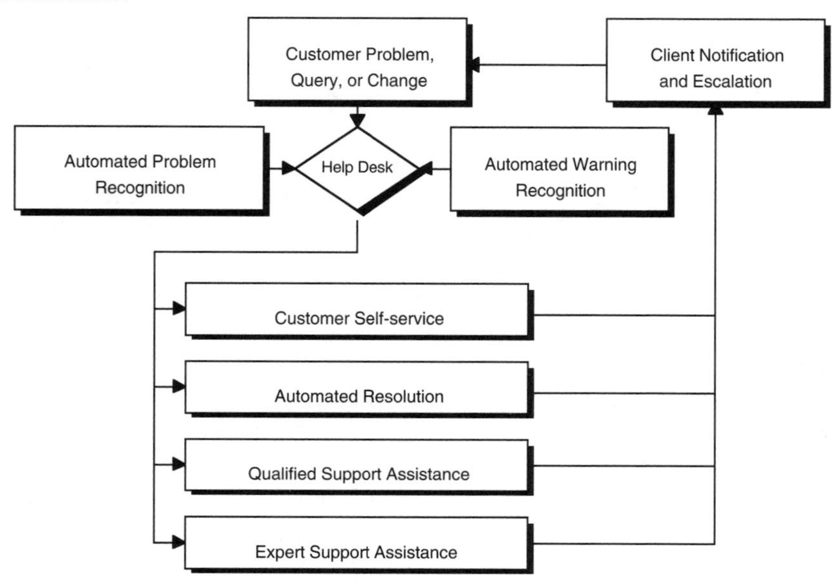

Figure 12.4 Flow Diagram of Client Services.

A target process was defined as providing a single-source help desk ("one stop shopping" approach) for ITO customers with performance targets as follows:

- 90 percent of calls to be answered by five rings.
- Less than 2 percent abandonment rate.

Other goals of the target process were to enhance the perception of the help desk and to significantly reduce the time required to resolve a customer's request. A combination of radical redesign ideas (reengineering) and incremental changes ideas (TQM) formed the nucleus of a target help desk process. The redesigned process implemented the following changes:

- Consolidate three help desks into one central help desk.
- Create a consistent means of problem/request logging and resolution.
- Introduce automation for receiving, queuing, and categorizing calls for resetting terminals.
- Capture information pertaining to the call once at the source and, if the call is handed off, have the information passed along also.
- Use existing technologies to create a hierarchy of problem resolution where approximately 60 percent of problems can be resolved immediately without using the operators and approximately 15 percent of the calls can be resolved immediately by the operators.
- Create an automated warning and problem recognition system that catches and deals with mainframe problems before they occur.

The original simulation model was revisited to better understand the current customer service level and what potential impact software changes, automation and consolidation would have on the staffing and equipment needs and operational capacity. Simulation results could also be used to manage the expectations for potential outcomes of the target process implementation.

Immediate benefit was gained from the use of this application of simulation to better understand the old operational interrelationships between staffing, call volume and customer service. Figure 12.5 shows how much the abandonment rate will change when the average daily call volume or the number of operators varies. The importance of this graph is realized when one notices that it becomes increasingly harder to lower the abandonment rate once the number of operators increases above seven. Above this point, the help desk can easily handle substantial increases in average daily call volume while maintaining approximately the same abandonment rate.

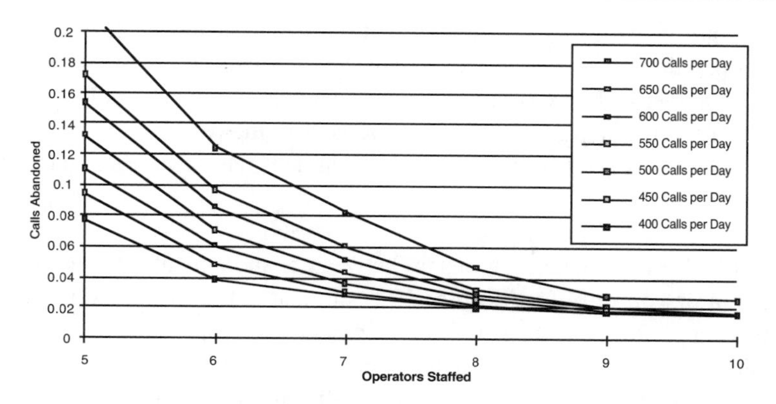

Figure 12.5 Abandonment Rates for Alternative Staff Levels.

After modeling and analyzing the current process, the project team evaluated the following operational alternatives using the simulation model:

- The option to select from a number of different shift schedules so that staffing can easily be varied from current levels.
- The introduction of the automated voice response unit and its ability to both receive and place calls automatically.
- The ability of the automated voice response unit to handle device resets, password resets, and system inquiries.
- The incorporation of the PC and LAN help desks so that clients with PC- related problems can have their calls routed directly to an available expert via the automated voice response unit.
- The ability to change the response time of ASIM problem logging system.

Additionally, two alternative staffing schedules were proposed. The alternative schedules attempt to better match the time at which operators are available for answering calls to the time the calls are arriving. The two alternative schedules reduce effort hours by up to eight percent while maintaining current service levels. Additional results related to the alternative operations simulation model were:

- The automated voice response unit will permit approximately 75 percent of PC-related calls to be immediately answered by a PC expert directly.
- Using Figure 12.5, the automated voice response unit's ability to aid in reducing the abandonment rate can be ascertained simply by estimating the reduction in the number of calls routed to help desk operators and finding the appropriate point on the chart for a given number of operators.

- Improving the response time of ASIM will have a noticeable effect on the operation when staffing levels are low and call volumes are high. For example, with five operators on staff and an average call volume of 650 calls per day, a 25 percent improvement in the response time of ASIM resulted in a reduction in the abandonment rate of approximately two percent.

Results: The non-linear relationship between abandonment rate and number of operators on duty (see Figure 12.5) indicates the difficulty in greatly improving performance once the abandonment rate drops below five percent. Results generated from the validated simulation model compare the impact of the proposed staffing changes with that of the current staffing levels. In addition, the analysis of the effect of the automated voice response unit can be predicted before implementation so that the best alternative can be identified.

The introduction of simulation to help desk operations has shown that it can be a powerful and effective management tool that should be utilized to better achieve operational goals and to understand the impact of changes. As the automation project continues to be implemented, the simulation model can greatly aid management and the project team members by allowing them to intelligently predict how each new phase will affect the help desk.

APPLICATION 5: DESIGN OF A FAST-FOOD RESTAURANT

Industry: Food Services (Aran and Kang 1987).

Background: The dynamic nature of the restaurant industry is ensured by the industry's goals of increased sales, market share, and profits. These goals continually lead to the introduction of new products, new marketing campaigns, new equipment, new ways of serving customers, and new facility designs. Burger King has used computer simulation in various applications.

Purpose of the Simulation Study: The models built at Burger King have been used to:

- Establish staffing levels.
- Design new restaurant facilities.
- Estimate impacts of equipment/procedural changes.

Model Description: A typical fast food restaurant has five sub-systems working within the overall restaurant system. These are:

1. Front counter customer
2. Drive-thru customer
3. Front-counter service
4. Drive-thru service
5. Kitchen

The service process is initiated by customers requesting orders either at the front counter or at the drive-thru speaker. They are serviced by the order takers, who in turn wait for the kitchen to fill the order. Customer interarrival times have been observed to be exponentially distributed, with the mean arrival time depending on the total sales generated through that service channel during a particular period.

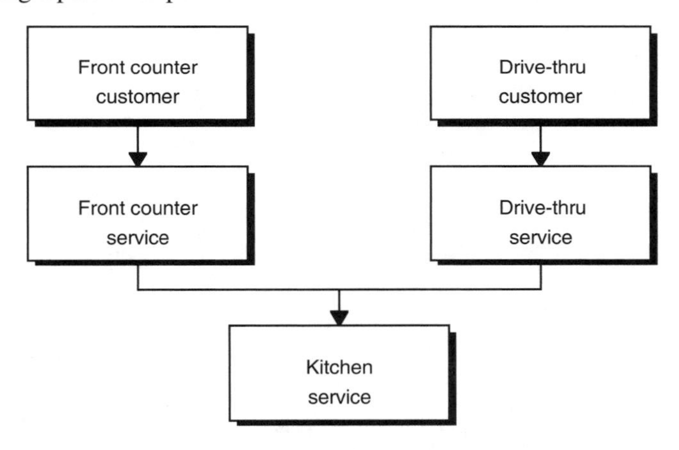

Figure 12.6 Flow Diagram of Order Operations.

When the customers arrive at the restaurant, they place their order, pay, and have their change made for them by the service crew members. The actual food items ordered by the customers must be determined and included in the model. Thus, order configuration is an important aspect of the model. Sampling customer tickets and coding them is one way of determining the frequency of each order configuration.

Once the customer has placed his/her order with the service crew member, this triggers and creates the demand for service from the kitchen sub-system. With this type of a "double" service system, fast and consistent service depends on the "right" balance between the service sub-systems and the kitchen sub-system. Some restaurants have been successful in simplifying this double service system by cooking food items ahead of time. However, at Burger King the "Have It Your Way" concept is the key to the marketing strategy and positioning. As such, Burger King uses the double service system.

Once the order is sent to the kitchen, a kitchen crewmember is assigned to prepare the order. Different types of orders are assigned to the various crew members. This assignment needs to be flexible in the model because during low levels only one person is needed in the kitchen; whereas during high levels of sales there may be three crew members in the kitchen. This flexibility in work assignments facilitates the use of one general model for various kitchen layouts and sales levels.

When running the full Burger King restaurant model and evaluating operational changes, the following system performance measures are considered:

- Customer total service time
- Customer line time to order
- Customer order time
- Customer change time
- Customer wait time for order
- Utilization of crewmembers
- Utilization of equipment

Results: As in many cases, a simulation model is used to determine an optimum configuration that achieves a particular objective function. For example, Burger King also used the model to arrive at an "Optimal Seating Configuration" for its restaurants. Specifically, how many tables for groups of two customers, groups of four customers, etc. should be used in order to maximize seating capacity. The results of this work were implemented in restaurants opened or remodeled in the last two years. The estimates indicate that the new designs have improved seating capacity by almost 15 percent.

Appendix A
Example of Input Data Collection

DATA COLLECTION FOR A FAST FOOD RESTAURANT

This example shows how data might be organized for modeling a service operation. The facility is a fast food restaurant shown below (Figure A.1).

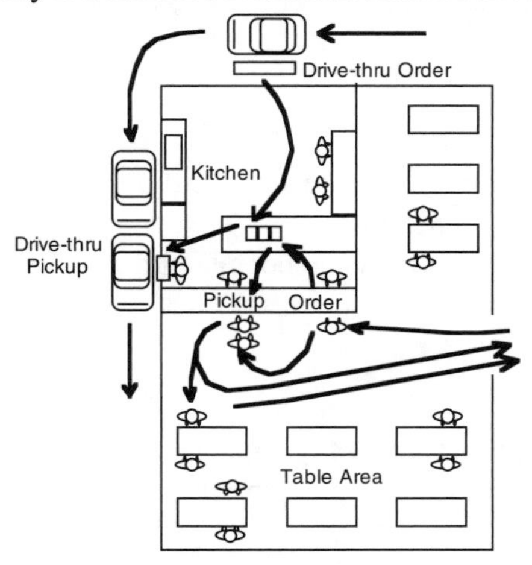

Figure A.1 Model of a Fast Food Restaurant.

Walk-in customers enter and place their order which is sent to the kitchen. They wait in the pickup queue while their order is being filled in the kitchen. Once they pick up their order, 60 percent of the customers stay and eat at a table, while the other 40 percent take out their order.

In addition to walk-in customers, there are also drive-thru customers that place their order and then wait in line to pick up their order.

If a walk-in customer enters and sees that their are more than 15 customers waiting to place their order, the customer will balk (leave). Drive-thru customers will also balk if there are more than five cars waiting to place an order.

The objective of the simulation is to analyze performance for different periods during the day to see how long customers spend waiting in line, how long lines become, how often customers balk, and what the utilization of the table area is.

The data has been organized into the following data modules (note that some data areas are shown for documentation purposes even if they are not used).

Entities

There are three entities included in the model. Since filled orders need to match up with the customer that placed the order, a *customer number* attribute is used.

Table A.1 Entities and Attributes for a Fast Food Restaurant.

Entity ID	Attributes
Walk-in customer	Customer order number
Drive-thru customer	Customer order number
Order ticket	Customer order number

Route Locations

Table A.2 shows:

- The different stop points or route locations where entities are either serviced or waiting to be serviced.
- The capacity or maximum number of entities that can occupy each location at any one time.

Table A.2 Location ID and Entity Capacity for Fast Food Restaurant.

Location ID	Entity Capacity
Customer order queue	15
Customer pickup queue	Unlimited
Drive-thru order queue	5
Drive-thru pickup queue	3
Table area	36
Order ticket queue	Unlimited

Resources

The resources consist of the following four types with each type dedicated to performing a specific activity.

Resource ID

- Order taker
- Order server
- Drive-thru server
- Kitchen crew

Processing Sequence

The following processing table (Table A.3) shows the routing sequence, processing times and other processing logic occurring at each location. With the limited sample times that were available, all processing times were assumed to follow a triangular distribution designated by T(min, mode, max).

Table A.3 Routing Sequence, Processing Times, and Other Processing Logic for Locations in a Fast Food Restaurant.

Entity	Location	Activity	Next Location
Walk-in Customer	Order Queue	Use Order taker T(.7, .9, 1.4) min. Assign a customer order number Create an Order-ticket for the Kitchen	Pickup Queue
	Pickup Queue	Match up with order Use Server for T(.5, .7, 1.1) min.	60% to table 40% exit
	Table Area	Stay for T(14, 21, 35) min.	Exit
Drive-thru Customer	Drive-thru Order Queue	Use Drive-thru server for T(.7, 1.2, 1.7) min. Assign a customer order number Create an Order-ticket for the Kitchen	Drive-thru Pickup Queue
	Drive-thru Pickup Queue	Match up with order Use Drive-thru server for T(.7, 1.2, 1.7) min.	Exit
Order-ticket	Order Ticket Queue	Use Kitchen crew for T(.8, 2.2, 4.5) Signal completion for this customer order number	Exit

Arrival Data

All customer arrivals are random (Poisson distributed) with both the arrival rate and mix of customers (walk-in and drive-thru) varying depending on the time of the day. The following table (Table A.4) identifies the customer mix and arrival rate for each period of the day.

Table A.4 Customer Mix and Arrival Rate in a Fast Food Restaurant.

Period of the day	Number of Arrivals per hr	% Walk-in Customers	% Drive-thru Customers
7-9 a.m.	15	50	50
9-11 a.m.	9	70	30
11 a.m. - 2 p.m.	26	70	30
2-5 p.m.	13	70	30
5-9 p.m.	32	75	25

Location and Resource Scheduling

All locations are available for 14 hours a day, 7 days a week. For the servers, there is a schedule that is prepared by the restaurant manager. Table A.5 shows the scheduled availability of server resources.

Table A.5 Scheduled Availability of Server Resources in a Fast Food Restaurant.

	Order takers on duty	Order servers on duty	Kitchen crews on duty
7-9 a.m.	6	6	6
9-11 a.m.	3	3	3
11 a.m. - 2 p.m.	7	7	7
2-5 p.m.	3	3	3
5-9 p.m.	6	6	6

Assumption List

Note that the following assumptions are being made in the routing:

- Move times are insignificant.
- All planned cleaning and maintenance are done during off-shift hours; therefore, they are not modeled.

Appendix B
Standard Theoretical Probability Distributions

This appendix describes common standard theoretical probability distributions that have applications in simulation. Most of these distributions are built into simulation software and only the defining parameters need be specified.

BERNOULLI DISTRIBUTION

The Bernoulli distribution (or Bernoulli trial) is a discrete distribution that applies to situations where there are two possible states (reject, non-reject). The probability of one state occurring is p; the probability of the other state occurring is $1 - p$. Phenomena that could be defined with a Bernoulli distribution include:

- The output of a process is either defective or non-defective.
- An employee shows up for work or not.
- An operation will require a secondary process or not.

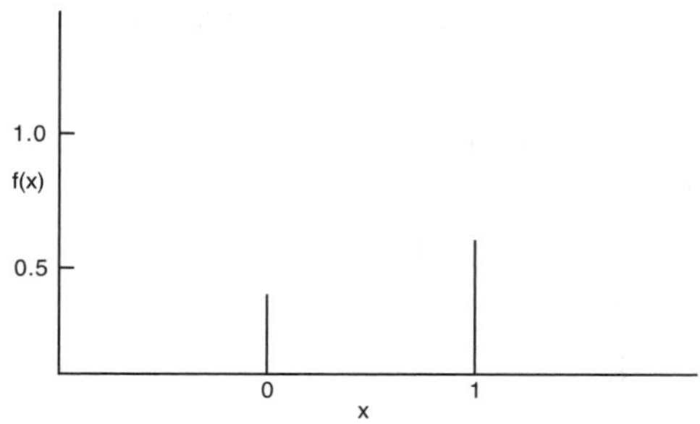

Figure B.1 Bernoulli Distribution with p = 0.60.

BINOMIAL DISTRIBUTION

The binomial distribution is a discrete distribution that expresses the number of outcomes in n trials. It is essentially the sum of n Bernouli trials. A binomial distribution is defined by the probability (p) of a particular outcome occurring in a number (n) of trials. Examples of phenomena that could be defined with a binomial distribution include:

- The number of defective items in a batch.
- The number of customers of a particular type that enter the system.
- The number of employees out of a group of employees who call in sick on a given day.

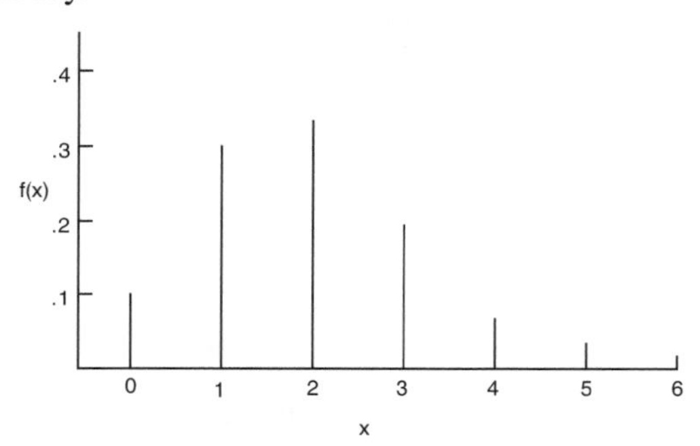

Figure B.2 Binomial Distribution with p = 0.30 and n = 6.

POISSON DISTRIBUTION

The Poisson distribution is a discrete distribution used to model a rate of occurrences (the number of occurrences of an event or characteristic per interval of time or per entity unit). It is used to define arrival rates, particularly in service systems. In a Poisson distribution, the mean is equal to the variance and is defined only by the average rate of occurrence or λ. Examples of phenomena that could be defined with a Poisson distribution include:

- The number of entities arriving each hour.
- The number of defects per item.
- The number of times a resource is interrupted each hour.

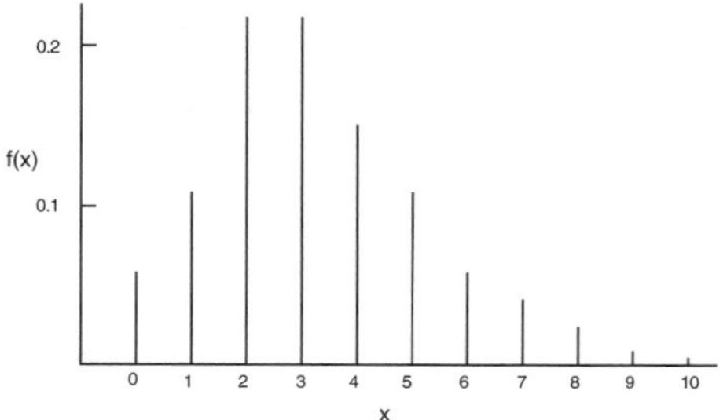

Figure B. 3 Poisson Distribution with λ = 4.0.

GEOMETRIC DISTRIBUTION

The geometric distribution is a discrete distribution that defines the number of trials before a particular outcome occurs. A geometric distribution is defined by identifying the probability (p) of any potential trial producing the outcome of interest. Examples of phenomena that could be defined using a geometric distribution include:

- The number of machine cycles before a failure occurs.
- The number of items inspected before a defective item is found.
- The number of customers processed before a customer of a particular type is encountered.

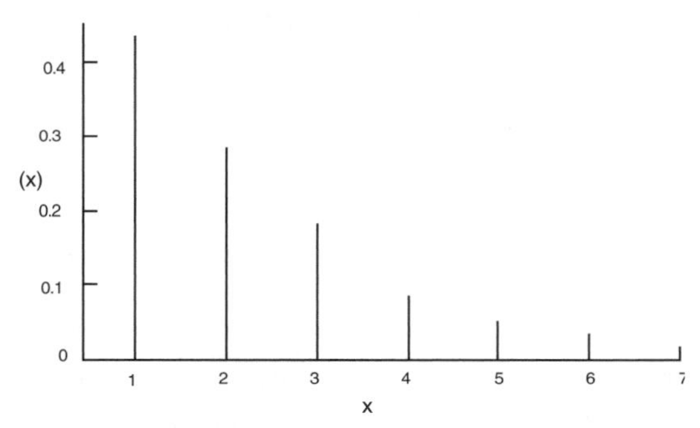

Figure B.4 Geometric Distribution with $p = 0.5$.

UNIFORM DISTRIBUTION

The uniform distribution is a continuous distribution used to define a value that is equally likely to fall anywhere within a specified range. Uniform distributions are defined by specifying the lower limit (a) and the upper limit (b) of the range. Alternatively, it is defined by specifying the midpoint and half range. The uniform distribution is seldom a valid representation of phenomena that occur in manufacturing and service systems. It is sometimes used when the underlying distribution is unknown.

Values from a uniform distribution may be truncated to provide a discrete uniform distribution (in this case, the upper limit specified for the distribution should be one higher than the highest value returned since $a \geq x > b$). An example of a phenomena that could be defined with a discrete Uniform distribution is:

- The type of an incoming entity given that each possible type is equally likely to occur.

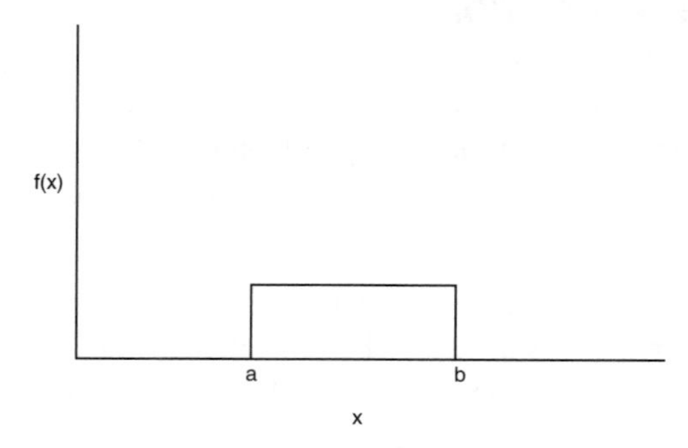

Figure B.5 Uniform Distribution.

TRIANGULAR DISTRIBUTION

The triangular distribution is easily understood and definable and provides a good first approximation of the true underlying distribution when data is sparse and no distribution fitting analysis has been performed. The triangular distribution is defined by the minimum value (a), a maximum value (b), and a mode (m). The triangular distribution is used to define activity times that tend to have a minimum, maximum, and most likely value. The weakness of the triangular distribution is that values in real activity times rarely taper linearly. Extreme values that are rare are also not captured by a triangular distribution.

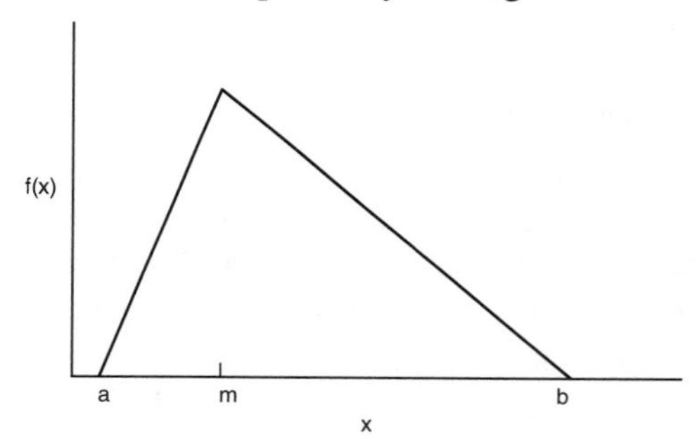

Figure B.6 Triangular Distribution.

NORMAL DISTRIBUTION

The normal distribution is a familiar distribution and relatively easy to understand and define. The normal distribution is defined by a mean (μ) and standard deviation (σ). The normal distribution is used to define activity times that tend to be symetrical about a central tendency (hence the bell shape). Contrary to what might at first seem to be the case, there are very few activities whose times fit a normal distribution. This is because most activities tend to be skewed (stretched out) to the right so that times are often quite longer than they are shorter from the mean.

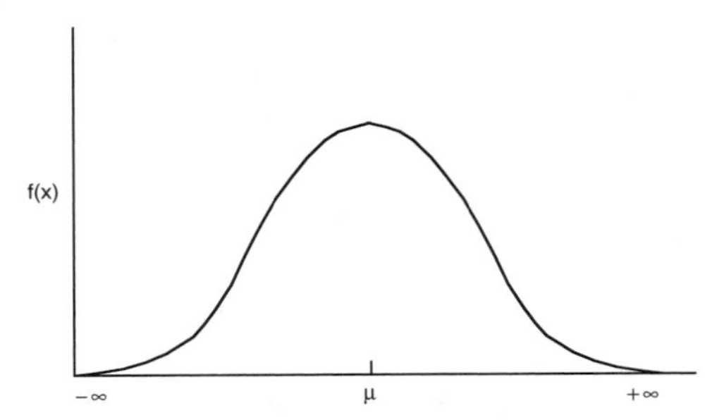

Figure B.7 Normal Distribution.

EXPONENTIAL DISTRIBUTION

The exponential distribution, sometimes called the negative exponential distribution, is closely related to the Poisson distribution in that, if an occurrence happens at a rate that is Poisson distributed, the time between occurrences is exponentially distributed. The exponential distribution is defined by the mean time (μ). The exponential distribution has a memory-less property that allows completely random occurrences to take place. The exponential distribution is used primarily to define intervals between occurrences such as the time between customer arrivals. Some activity times such as repair times or the duration of telephone conversations may also be exponentially distributed.

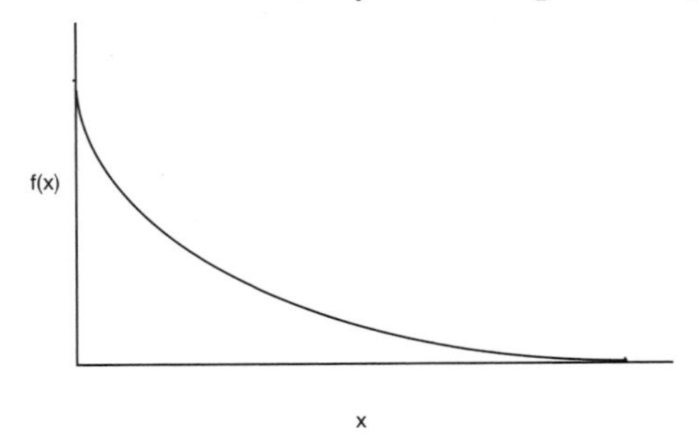

Figure B.8 Exponential Distribution.

BETA DISTRIBUTION

The beta distribution is used when insufficient data is available to define a more accurate distribution. It defines random proportions such as the percentage of defective items in a lot, and activity times, particularly when multiple tasks make up the activity as in a project (PERT network activity times assume a beta distribution). The beta distribution is defined by two shape parameters and must be scaled to provide ranges greater than one.

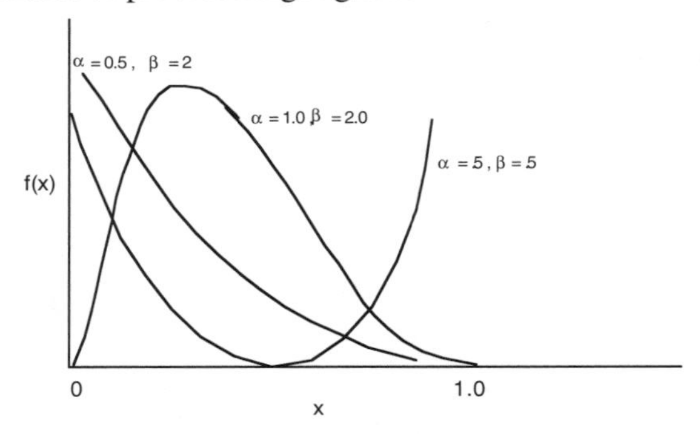

Figure B. 9 Beta Distribution.

LOGNORMAL DISTRIBUTION

The lognormal distribution is a useful distribution that is characteristic of activities having multiple subactivities. It is often used to define manual activities such as assembly, inspection or repair. The time between failures is often lognormally distributed.

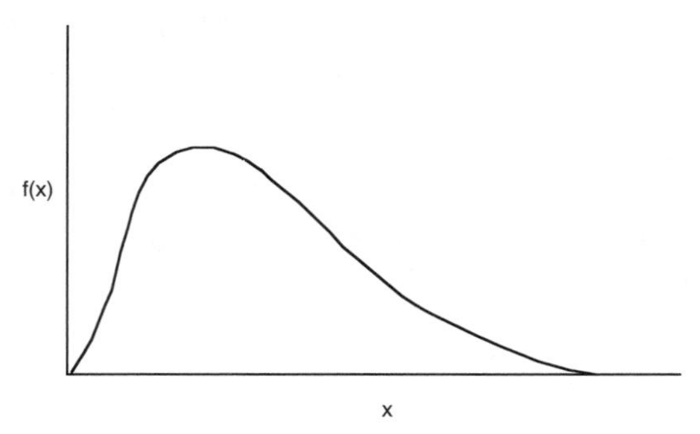

Figure B.10 Lognormal Distribution.

GAMMA DISTRIBUTION

The gamma distribution takes on a variety of different shapes and is therefore a useful distribution for describing time values for many different types of activities. Typical activities whose time may be defined by a gamma distribution include manual tasks such as service times or repair times. A special case of the gamma distribution is the erlang.

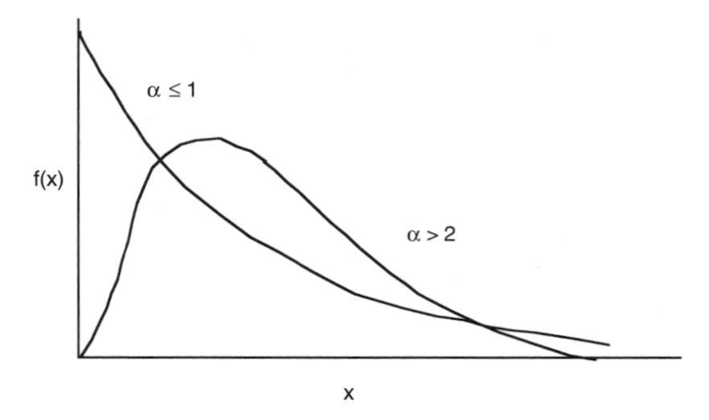

Figure B.11 Gamma Distribution.

WEIBULL DISTRIBUTION

The Weibull distribution is similar to the gamma distribution, having many different shapes. It is often used in reliability theory for defining the time until failure particularly due to items (bearings, tooling, etc.) that wear.

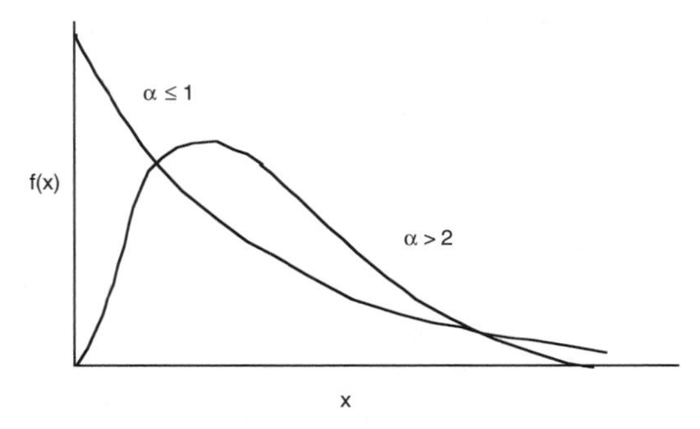

Figure B.12 Weibull Distribution.

Appendix C
Confidence Interval Estimation

To illustrate the use of a confidence interval, suppose that 10 replications of an experiment were performed to test the effect of a certain set of conditions on the expected waiting time of customers in a queue. Assume, further, that the average waiting time for all replications is 21.5 minutes. To establish a 95 percent confidence interval, only the half-width value (*hw*) would need to be calculated. Then we could say that there is a 95 percent chance (or a .95 probability) that the true mean lies inside the interval from 21.5 – *hw* to 21.5 + *hw*. This is another way of saying that we can be 95 percent confident that our sample mean $\overline{X}$ of 21.5 is within ± *hw* of the true mean (μ) which is unknown. If we are not satisfied with the interval and would like it to be smaller, we can always take more samples. Taking more samples can also increase our confidence for a particular interval.

To calculate the half-width of a confidence interval for a given probability or level of confidence, the following equation is used:

$$hw = (Q)(s / \sqrt{n})$$

where Q is a factor (called a quantile) that can be looked up in a statistical table called a *Student t* table.

Quantile values in the *Student t* table are identified according to the value of $(1 - \alpha/2)$ and the degrees of freedom (n - 1). The term α is the complement of P (1 - P), and is referred to as the *significance level.* The significance level may be thought of as the "risk level" or the probability that μ will fall outside of the confidence interval. To find the Q value for a .90 confidence level for a sample size of 20, for example, we would look under the $1 - \alpha/2$ value of .95 and the degrees of freedom of 20 –1 or 19. The resulting Q value is 1.729.

To show a complete example of how a confidence interval is established, suppose that five replications are made with the following waiting times reported for each replication:

Replication	**Waiting Time (minutes)**
1	44
2	36
3	38
4	47
5	39

If X_i is the waiting time value reported by the *i*th replication, true expected waiting time can be estimated by calculating $\overline{X}$ as follows:

$$\overline{X} = \frac{\sum_{i=1}^{5} x_i}{5}$$

$$= 40.80 \text{ minutes}$$

our sample size standard deviation (s) is

$$s = \sqrt{\frac{\sum_{i=1}^{5} [X_i - 40.8)]^2}{4}}$$

$$= 4.55$$

Using a *Student t* table, the quantile for a .95 confidence level and n – 1 or 4 degrees of freedom is 2.776. The following calculates the half-width (*hw*):

$$\text{hw} = 2.776\left(\frac{4.55}{\sqrt{4}}\right)$$

$$= 6.315$$

Since the confidence interval is expressed as $\overline{X} \pm$ hw, the lower limit and upper limits of the interval are calculated as follows:

Lower limit $= \overline{X} - hw$
$= 40.80 - 6.315$
$= 34.48$ minutes

Upper limit $= \overline{X} + hw$
$= 40.80 + 6.315$
$= 37.12$ minutes

So we can say that we are 95 percent confident that the true mean (μ) falls between 34.48 minutes and 47.12 minutes.

References

Adams, W. 1985. Can AGVSs Help to Integrate Your Factory Operations? *CIM Review*, Fall, 50-57.

Aran, M.M., and Kang, K. 1987. Design of a Fast Food Restaurant Simulation Model. *Simulation.* Norcross, GA: Industrial Engineering and Management Press.

Askin, R. G., and Standridge, C. R. 1993. *Modeling and Analysis of Manufacturing Systems*. New York: John Wiley & Sons.

Banks, J., and Carson, J. S. II. 1984. *Discrete-Event System Simulation*. Englewood Cliffs, NJ: Prentice Hall.

Bevans, J. P. 1982. First, choose an FMS simulator. *American Machinist*. May 1982: 144–5.

Blackburn, J., and Millen, R. 1986. Perspectives on Flexibility in Manufacturing: Hardware Versus Software. *Modeling and Design of Flexible Manufacturing Systems*, ed. Andrew Kusiak. Amsterdam: Elsevier.

Blackburn, J.D. 1992. Time-Based Competition: White-Collar Activities. *Business Horizons* 35(4): 97–8.

Blanchard, B. S. 1991. *System Engineering Management*. New York: John Wiley & Sons.

Boehlert, G., and Trybula, W. J. 1984. Successful Factory Automation. *IEEE Transaction*, September 1984: 218.

Bozer, Y. A., and White, J. A. 1984. Travel-Time Models for Automated Storage/Retrieval Systems. *IIE Transactions*. 16(4): 329–38.

Brooks, F. P. Jr. 1978. *The Mythical Man-Month*. Reading, MA: Addison-Wesley,

Browne, J. 1994. Analyzing the Dynamics of Supply and Demand for Goods and Services. *Industrial Engineering*. 26(6): 18–19.

Buzacott, J. A. and Shanthikumar, J. G., 1993. *Stochastic Models of Manufacturing*

Systems. Englewood Cliffs, N.J.: Prentice Hall.

Carrie, A. 1988. *Simulation of Manufacturing Systems*. New York: John Wiley & Sons.

Carson, J. S. 1986. Convincing User's of Model's Validity is Challenging Aspect of Modeler's Job. *Industrial Engineering*. June 1986:77.

Charles Stark Draper Lab. 1983. *Flexible Manufacturing System Handbook, Vol II*. Cambridge, MA.

Charles Stark Draper Lab. 1983. *Flexible Manufacturing System Handbook, Vol. V*. Cambridge, MA.

Christy, D. P., and Watson, H. J. 1983. The Application of Simulation: A Survey of Industrial Practice. *Interfaces*, 13(5).

Collier, D. A. 1994. *The Service/Quality Solution*. Milwaukee: ASQC Quality Press.

Conway, R., and Maxwell, W. L. 1986. XCELL: A Cellular, Graphical Factory Modelling System. In *Proceedings of the 1986 Winter Simulation Conference*. J. Wilson, J. Henriksen, and S. Roberts, eds. p. 160.

Conway, R.; Maxwell, W. L.; and Worona, S. L.1986.*User's Guide to XCELL Factory Modeling System*. Palo Alto: The Scientific Press.

Cutkosky, M.; Fussel, P.; and Milligan, R. Jr. 1984. The Design of a Flexible Machining Cell for Small Batch Production. *Journal of Manufacturing Systems*. 3(1): 39.

Davenport, T. H. 1993. *Process Innovation*. Cambridge, MA: Harvard Business School Press.

Deming, W. E. 1989. Foundation For Management of Quality in the Western World. Paper read at a meeting of the Institute of Management Sciences, 24 July 1989, Osaka, Japan.

Fitzgerald, K. R., 1985. How to Estimate the Number of AGVs You Need. *Modern Materials Handling*. 40(12): 79

Gershwin, S. B. 1994. *Manufacturing Systems Engineering*, Englewood Cliffs, NJ: Prentice Hall.

Glenney, N. E., and Mackulak, G. T., 1985. Modeling and Simulation Provide Key to CIM Implementation Philosophy. *Industrial Engineering*. 17(5): 76.

Gordon, G. 1978. *System Simulation*, 2d ed. Englewood Cliffs, NJ: Prentice-Hall.

Gottfried, B. S. 1984. *Elements of Stochastic Process Simulation*. Englewood Cliffs, N.J.: Prentice-Hall.

Greene, J. H. 1987. *Production and Inventory Control Handbook*. New York: McGraw-Hill.

Guerrera C., and Davis K. 1992. Optimizing Picking Methods in a Warehouse. In *Proceedings of the August 1992 PROMODEL Users Conference*.

Hammer, M., and Champy, J. 1993. *Reengineering the Corporation*. New York: Harper-Collins.

Hancock, W.; Dissen, R.; and Merten, A. 1977. An Example of Simulation to Improve Plant Productivity. *AIIE Transactions*. 9(1): 2-10.

Hancock, W.; Magerlein, D.C.; Storer, R.H.; and Martin, J.B. 1978. Parameters Affecting Hospital Occupancy and Implications for Facility Sizing. *Health Services Research*. 13: 276-289.

Harrell, C.R. 1993. Modeling Beverage Processing Using Discrete Event Simulation. In *Proceedings of the 1993 Winter Simulation Conference*. G. Evans, M. Mollaghasemi, E. Russell, and W. Biles eds.

Harrell, C. R.; Bateman, R. E.; Gogg, T.; and Mott, J. R. A. 1992. *System Improvement Using Simulation*. Orem, UT: PROMODEL Corporation.

Harrington, J.H. 1991. *Business Process Improvement: The Breakthrough Strategy for Total Quality, Productivity and Competitiveness*, New York: McGraw-Hill.

Hein G., and Jones T. 1991. Inventory, Manufacturing Operations and Customer Service Simulation Model. In *Supplementary Proceedings of the 1991 Autofact Conference.*

Henderson, R., and Kiran, A. 1992. Minimum Cost Test Planning. In *Proceedings of Nepcon West 1992 Conference.*

Henriksen, J., and Schriber, T. 1986. Simplified Approaches to Modeling Accumulation and Non-Accumulating Conveyor Systems. In *Proceedings of the 1986 Winter Simulation Conference.* J. Wilson, J. Henricksen, and S. Roberts, eds.

Hollier, R. H. 1980. The Grouping Concept in Manufacture. *International Journal of Operations Prod. Management.* 1:71.

Hoover, S. V., and Perry, R. F. 1990. *Simulation: A Problem Solving Approach.* Reading MA: Addison-Wesley.

Kharwot, A.K. 1991. Computer Simulation: An Important Tool in the Fast-Food Industry. In *Proceedings of the 1991 Winter Simulation Conference*. B. Nelson; W. D. Kelton; and G.M. Clark, eds.

Kiran, A. S.; Roberts, D.; and Strudler S. 1993. A Simulated SMT Production Line. *Surface Mount Technology*. August 1993: 27–29.

Knepell, P. L., and Arangno, D. C. 1993. *Simulation Validation*. Los Alamitos: IEEE Computer Society Press.

Kochan, D. 1986. *CAM Developments in Computer Integrated Manufacturing*. Berlin, Heidelberg: Springer-Verlag.

Kraitsik M.J., and Bossmeyer A. 1992. Simulation Applied to Planning an Emergency Department Expansion. In *Proceedings of the February 1992 SCS Western Multiconference.*

Law, A. M. 1986. Introduction to Simulation: A Powerful Tool for Analyzing Complex Manufacturing Systems. *Industrial Engineering*. 18(5): 57–8.

Law, A.M.1991. Designing and Analyzing Simulation Experiments. *Industrial Engineering*. 23(3): 20-23

Law, A. M., and Kelton, D. W. 1991. *Simulation Modeling and Analysis*. New York: McGraw-Hill.

Law, A. M., and McComas, M. G. 1988. How Simulation Pays Off. *Manufacturing Engineering*. 100(2): 37-39.

Lewis, C. 1993. Cellular Design for Ophthalmic Lens Manufacturing. In *Proceedings of PROMODEL Users Conference.* August 1993.

Little, J.D.C. 1961. A Proof for the Queuing Formula: $L = \lambda W$. *Operations Research*. 9(3): 383–7.

Mabrouk, K. 1993. A Step-by-Step Project for Designing an Assembly and Test System. In *Proceedings of the 1993 Autofact Conference.*

Marmon, C. 1991. Teledyne Applies Simulation to the Design and Justification of a New Facility. *Industrial Engineering*. March 1991: 29-32

Mauer, J.L., and Schelasin, R.E. The Simulation of Integrated Tool Performance in Semiconductor Manufacturing. In *Proceedings of the 1993 Winter Simulation Conference.* G. Evans, M. Mollaghasemi, E. Russell, and W. Biles eds.

Meyers, F. E. 1993. *Plant Layout and Material Handling*. Englewood Cliffs, NJ:

Regents/Prentice Hall.

Miner, R. J. 1987. Simulation as a CIM Planning Tool. *Simulation.* R. D. Hurrion, ed. UK: IFS Publications.

Mitra, D., and Mitrani, I. 1990. "Analysis of a Kanban Dicipline for Cell Coordination in Production Lines," *Management Science*. 36(12): 1548-66.

Nadler, G. 1965. What Systems Really Are. *Modern Materials Handling*. 20 (7):41–7.

Nedza, P. 1994. Simulation As An Operations Management Tool. In *Proceedings of the April 1994 International Integrated Manufacturing Conference.*

Neelamkavil, F. 1987.*Computer Simulation and Modelling.* New York: John Wiley & Sons.

Nixon, E.1994. Mass Production: Financial Services. In *Proceedings of the February 1994 PROMODEL Midwest Users Conference.*

Noaker, Paula M. 1993. Strategic Cell Design. *Manufacturing Engineering.* March 1993: 81-84..

Perry, R. F., and Baum, R. F. 1976. Resource allocation and scheduling for a radiology department. *Cost Control in Hospitals.* Ann Arbor: Health Administration Press.

Pritsker, A. A. B. 1986. *Introduction to Simulation and Slam II.* West Lafayette, IN: Systems Publishing Corporation.

Pritsker, A. B., and Pegden, C. D. 1979. *Introduction to Simulation and SLAM.* New York: John Wiley & Sons.

Reedy, W. 1993. Quantification and Improvement of Throughput Capabilities for Aircraft Maintenance Using Simulation Modeling. In *Proceedings of August 1993 PROMODEL Users Conference.*

Renbold, E. V, and Dillmann, R. 1986. *Computer-Aided Design and Manufacturing Methods and Tools.* Berlin, Heidelberg: Springer-Verlag.

Ronald, G. A. and Standridge, C. R., 1993. *Modeling and Analysis of Manufacturing Systems.* New York: John Wiley & Sons.

Ross, S. M. 1990. *A Course in Simulation.* New York: Macmillan.

Sasser W.E.; Olsen R.P.; and Wyckoff D.D. 1978. *Management of Service Operations.* Boston: Allyn & Bacon.

Schlesinger, S. 1979. Terminology for Model Credibility. *Simulation.* 32(3): 103-4.

Schmenner, R. W. 1994. *Plant and Service Tours in Operations Management*, 4th ed. New York: Macmillan.

Schonberger, R. J., and Knod, E. M., Jr. 1994. *Operations Management: Continuous Improvement*, 5th ed. Burr Ridge, IL: Irwin.

Schwind, G. F. 1988. Automatic Monorail Systems. *Material Handling Engineering.* May 1988: 95.

Shannon, R. E. 1975. *System Simulation: The Art and Science.* Englewood Cliffs, NJ: Prentice-Hall.

Shannon, R. et al. 1980. Operation research methodologies in industrial engineering: A Survey. *AIIE Transactions.* 12(4): 364.

Shingo, S. 1992. *The Shingo Production Management System-Improving Process Functions.* Translated by Andrew P. Dillon. Cambridge, MA: Productivity Press.

Singer R. and Gasparatos A. 1994. Help Desks Hear Voice. *Software Magazine.* February 1994: 24-26.

Solberg, J. 1988. *Design and Analysis of Integrated Manufacturing Systems.* Edited by W.

Dale Compton. Washington: National Academy Press.

Suri, R., and Tomsicek, M. 1988. Rapid Modeling Tools for Manufacturing Simulation and Analysis.In *Proceedings of the 1988 Winter Simulation Conference*. Abrams, M.A.; Haig, P.L; and Comfort, J.C.; eds. pp. 25-32.

Suzaki, K. 1987. *The New Manufacturing Challenge: Techniques for Continuous Improvement*. New York: The Free Press.

The Material Handling Institute. 1977. *Consideration for Planning and Installing an Automated Storage/Retrieval System.* Pittsburgh, PA.

Thesen, A., and Travis, L. E. 1992. *Simulation For Decision Making*. St. Paul: West Publishing Company.

Tompkins, J.A., and White, J.A. 1984. *Facilities Planning*. New York: John Wiley & Sons.

Toffler, A. 1980. *The Third Wave.* New York: Morrow.

Wang, H., and Wang, H. P. 1990. Determining the Number of Kanbans: A Step Toward Non-Stock Production. *International Journal of Production Research*. 28(11): 2101–15.

Whitworth, S. 1993. Simulating the Thrills and Drills of a Dental Clinic. In *Proceedings of the August 1993 PROMODEL Users Conference.*

Wick, C. 1987. Advances in Machining Centers. *Manufacturing Engineering*. 99(4): 24.

Widman, L. E.; Loparo, K. A.; and Nielsen, N. R. 1989. *Artificial Intelligence, Simulation, and Modeling*. New York: John Wiley & Sons.

Zilm, F., and Hollis R 1983. An Application of Simulation Modeling to Surgical Intensive Care Bed Need Analysis. *Hospital and Health Services Administration*, Sept/Oct. 82-101.

Index

–A–

Activities, 19
Aerospace simulation example, 238–40
Analytic models, 27–28
Animation processor, 43
Animation software, 63–64
Arrays, 113
Arrivals, entity, 108–9
Attributes, 95, 113
Automated storage/retrieval system (AS/RS)
  analytical cycle time calculations, 193–95
  carousels, 196–97
  computerized cycle time calculations, 193
  configuring, 192–93
  modeling, 195–96
  with picking operations, 195
Automatic guided vehicle system (AGVS), 197
  designing, 198–99
  managing, 199–200
  modeling, 200–201
  shuttle carts, 201–2
Automation
  manufacturing design and, 146–47
  service industry and, 214
Automotive simulation example, 240–42

–B–

Baker and Taylor Books, 257–59
Batch flow shop, 169–71
Batching process, manufacturing and, 140, 148–49
Batch means technique, 131
Bernoulli distribution, 287
Beta distribution, 292
Beverage industry, simulation example, 259–62
Binomial distribution, 288
Burger King, 280–82

–C–

Canadian Imperial Bank of Commerce (CIBC), 273–75
Capacity planning, manufacturing and, 152
Carousels, 196–97
Cause-and-effect relationships, 21–22, 79–80
CCH Incorp., 264–66
Cell management, 137–38
Cellular manufacturing, 161–65
Central limit theorem, 125

College–Industry Council on Material Handling Education (CICMHE), 180
Communication and sales, simulation used in, 10–11
Computer–integrated manufacturing (CIM), 147
Conditional events, 44–46, 79
Confidence interval estimation, 123, 294–96
Constructs. See Modeling/models
Consumable resources, 100
Continuous-change state variables, 35
Continuous process systems, 177–78
Continuous simulation, discrete-event versus, 34–35
Controlled variables, 88
Controls, 19–20
Conveyors
load control of, 184–85
modeling networks of, 188–89
modeling of, 186–87
modeling of single section, 187–88
operational characteristics, 185
types of, 182–83
Costs
of simulation, 13–15, 69–71
of software, 65–66, 70
Cranes and hoists, 202–3
Cyclic arrivals, 108, 109

–D–

Data
collection and analysis, 77–85
observational versus time-weighted, 117-19
Decision variables, 88
Delivery services, 230-32
Detailed histories, 117
Deterministic simulation, stochastic versus, 35-36
Direct-response relationships, 22
Discrete-change state variables, 35
Discrete-event simulation
how it works, 44–46
versus continuous simulation, 34–35
Distribution
empirical versus user-defined, 83
simulation example, 257–59
theoretical, 83, 287–93
Documentation
of implementation of simulation model, 91–92
software, 65
Downtimes and repairs, modeling, 110–13

–E–

Electronics simulation example, 252–54
Empirical distribution, 83
Emulation, 139
Entities/entity
arrivals, 108–9
processes, 105–6
role of, 18–19, 95–96
routings, 103–4
service industry, 211
Events, scheduled and conditional, 44–46, 79
Experimental variables, 88
Expert systems, 34
Exponential distribution, 291

–F–

Face validity, 88
Financial services, simulation examples, 273–80
Flexible manufacturing system (FMS), 165–68
Flow control, service industry and, 215
Flow diagrams, 26–27
Food services
data collection and, 283–86
simulation example, 280–82

–G–

Gamma distribution, 293

General-use resources, 99
GE Nuclear Energy, 14
Geometric distribution, 289
Graphics, 63–64
Group technology (GT), 161

–H–

Hardware requirements
  costs of, 71
  software selection and, 65
Healthcare services, simulation examples, 267–72
HEI Corp., 139
Hughes Network Systems, 252–54
Hypothesis testing, 132

–I–

IBM Credit Corp., 4
IMED Corp., 262–64
Implementation steps
  conduction of experiments, 88–91
  data collection and analysis, 77–85
  documentation and presentation of results, 91–92
  general procedure, 73–75
  model development, 85–88
  objectives/constraints identified, 75–77
Independent/input variables, 88
Industrial vehicles, 189–91
Input selection rules, 98
Interactive modeling, 10
Interfacing, external, 64
Internally initiated arrivals, 108, 109
Interval batching, 131
Inventory control, manufacturing and, 150–52
ISO 9000, 6

–J–

Job sequencing, 138
Job shop, 157–61
Just-in-time (JIT), 149–50

–K–

Kanban, 149–50

–L–

Languages, simulation, 39, 58
  constructs, 41–42, 62
Layout design
  manufacturing and, 145–46
  service industry and, 213–14
Life cycles
  changes in product/service, 2–3
  model, 86
Line flow manufacturing
  production/assembly lines, 171–74
  transfer lines, 174–77
Loads, 95
Local variables, 113
Logic, modeling, 113–14
Lognormal distribution, 292

–M–

Maintainability, 111, 141
Management commitment, 51–52
Manufacturing
  applications of simulation in, 135–37
  classifications of, 153–77
  decision horizon for, 137–38
  emulation, 139
  factors affecting modeling of, 144–53
  performance measures, 142–44
  software for, 59
  terminology for, 139–42
Manufacturing, examples of simulation in
  aerospace, 238–40
  automotive, 240–42
  beverage, 259–62
  consumer products, 246–52
  electronics, 252–54
  medical instruments, 262–64
  metal fabrication, 242–46
  publishing/printing, 264–66

semiconductor, 254–57
warehousing/distribution, 257–59
Manufacturing lead time (MLT), 142
Material handling
classifications of, 181–205
movement, 102–3
principles of, 179–81
resources, 100
software for, 59
Material requirements planning (MRP), 149, 151
Mean time between failures (MTBF), 111, 141
Mean time to repair (MTTR), 111, 141
Medical instruments, simulation example, 262–64
Metal fabrication, simulation example, 242–46
Modeling/models
analytic, 27–28
constructs, 41, 62
description of, 24–25
development of, 85–88, 93–114
elements for building, 94–95
entities, 95–96
entity arrivals, 108–9
entity processes, 105–6
entity routings, 103–4
fidelity, 85
flexibility, 62–63
interface module, 41–42
life cycle, 86
logic, 113–14
movement, 100–103
paradigms, 93–94
partitioning, 86
processing module, 42
resource availability schedules, 109
resource downtimes and repairs, 110–13
resources, 96–100
resource setups, 109–10
selection of, 30–31
services, 67
simulation, 29–30
symbolic, 26–27
verification and validation, 87–88
Move batch, 148
Movement, modeling, 100–103
Multiple replication summaries, 117
Multiple scenario summaries, 117

–N–

National Institute of Standards and Technology (NIST), 162
Nonterminating simulations. See Steady-state simulations
Normal distribution, 291

–O–

Objective function, 23
Observational data, time-weighted versus, 117–19
Optimization, 22–24
Output
interface module, 44
measures, 119–21
observational versus time-weighted, 117–19
processor, 43–44
queuing rules, 98–99
reports, types of, 116–17

statistical analysis of, 121–24
statistical problems with, 124–25

–P–

Paradigms, modeling, 93–94
Part family/cell layout, 145–46
Parts, 95
Path network movement, 102
Performance
factors affecting, 11–12
measures for manufacturing, 142–44
measures for service industry, 209–11
measures for systems, 20
need for improved, 3–4
trial-and-error methods used to improve, 4–5
variables, 88

Periodic arrivals, 108
Poisson distribution, 288
Policy decisions, service industry and, 214
Probabilistic outcomes, simulation of, 36–38
Process batch, 148
Process change studies, 138
Processes, systems versus, 18
Processing, service industry, 212
Process layout, 145–46
Production batch, 148
Production control, manufacturing and, 149–50
Production scheduling, manufacturing and, 138, 152–53
Product layout, 145–46
Products, 95
Professional services, 225–27
Project shop, 154–56
Public relations, simulation used in, 11
Publishing/printing, simulation example, 264–66
Pull system, 149
Pure service shop, 218–22
Push system, 149

–Q–

Queue-limiter, 150
Queuing rules, 98–99
Queuing systems, 39

–R–

Random number generator, 37
Random streams, 132–34
Random variates, 38
Rapid Modeling Technology (RMT), 28
Relative ranking, 68
Reliability, 141
Reorder point system, 151
Replicating the simulation, 128
Replications, running independent, 131
Resources
  availability, 110
  consumable, 100
  general-use, 99
  loading, 138
  manufacturing and, 152
  material handling, 100
  modeling availability schedules, 109
  modeling downtimes and repairs, 110–13
  modeling setups, 109–10
  role of, 19, 96
  route location, 96–99
  service industry, 211–12
Response variables, 88
Retail service stores, 222–25
Robots, 203–5
Route location resources, 96–99
Routings, entity, 103–4
Rule of tens, 8
Run length, determining, 131–32

–S–

Scheduled arrivals, 108
Scheduled events, 44–46, 79
Scheduling software, 59
Scott & White, 267–70
Semiconductor simulation example, 254–57
Service factory, 216–18
Service industry
  applications of simulation in, 208–9
  characteristics of, 207
  classifications of, 215–35
  factors affecting modeling of, 213–15
  general simulation procedures, 212–13
  modeling considerations, 211–12
  performance measures, 209–11
  software for, 59
Service industry, examples of simulation in
  financial services, 273–80
  food services, 280–82
  healthcare services, 267–72

Simulation
benefits of, 15–16
components of, 40–44
costs of, 13–15, 69–71
defined, 33–34
example of, 46–50
history of, 39–40
interface module, 42
models, 29–30
pitfalls/precautions, 15, 92
popularity of, 6–7
processor, 43
reasons for using, 5–6
steps for getting started, 52–58
steps for implementation, 73–92
types of, 34–36
uses of, 7–13
Simulators, 39, 58
Single run summaries, 117
Snapshot reports, 117
Society Bank, 275–80
Software
costs of, 65–66, 68
documentation, 65
ease-of-use, 60–62
evaluating, 59–67
graphics and animation, 63–64
hardware requirements, 65
support, 66
training, 66
types of, 58–59
upgrades and enhancements, 67
weighted score selection of, 67–69
SOLA, 250–52
Staffing, service industry and, 214–15
State variables, 42
Statistical analysis of output, 121–25
Statistical capabilities of software, 16, 64–65
Statistical counters, 42
Steady-state simulations
analysis of, 129–32
terminating versus, 126–28
Steady-state versus transient behavior, 125–26
Stochastic versus deterministic simulation, 35–36
Supervised variables, 88
Symbolic models, 26–27
System(s)
configuration, 138
defined, 17–18
elements of, 18–20
performance measures, 20
verses processes, 18
Systems analysis and design
challenges facing, 1–4
modeling and, 24–31
optimization approach to, 22–24
pyramid, 24
simulation used in, 7–9, 136, 208–9
systems approach to, 21–22
Systems approach, 21–22
Systems management, simulation used in, 9–10, 136–37, 209

–T–

Technology assessment, 138
Teledyne Allvac, 242–46
Telephonic services, 227–30
Terminating simulations
analysis of, 128–29
steady-state/nonterminating versus, 126–28
Theoretical distribution, 83, 287–93
Timeline for simulation preparation, 71–72
Time-weighted data, observational versus, 117–19
Training
costs of, 70
software and, 66
Training and education, simulation used in, 10
Transactions, 95
Transfer batch, 148
Transient behavior, steady-state versus, 125–26

Transportation services, 232–35
Triangular distribution, 290

–U–

Uniform distribution, 289
User-defined distribution, 83

–V–

Variables, 113
  independent/input, 88
Visual interactive simulation, 40

–W–

Warehousing/distribution,
  simulation example, 257–59
Warm-up period, 129–30
Weibull distribution, 293
Work-in-process (WIP), 143
Workstation design, service industry and, 214

About the Authors

Charles R. Harrell is an associate professor of Manufacturing Engineering at Brigham Young University, and founder and chairman of PROMODEL Corporation, in Orem, Utah. Dr. Harrell received his B.S. in Manufacturing Engineering Technology from Brigham Young University; M.S. in Industrial Engineering from University of Utah; and Ph.D. in Manufacturing Engineering from the Technical University of Denmark. Prior to forming PROMODEL, he worked in simulation and systems design for Ford Motor Company and Eaton Kenway Corporation. Dr. Harrell is a senior member of IIE and SME.

Kerim Tumay is the Director of Process SImulation for CACI Products Company. Tumay received his B.S. and M.S. degrees in Industrial Engineering from Arizona State University. Prior to joining CACI, Tumay was Vice President of marketing with PROMODEL Corporation. Since 1983, Tumay has provided simulation training and modeling services to many organizations including Baxter, DuPont, Digital, IBM, Motorola, and Philips. He has published more than 40 papers on production and business process modeling and chaired simulation sessions at major conferences such as Autofact, Nepcon, National BPR Conference, and the Winter Simulation Conference.